THE SEVEN PILLARS OF
STATISTICAL WISDOM

THE SEVEN PILLARS OF
STATISTICAL
WISDOM

STEPHEN M. STIGLER

Harvard University Press

Cambridge, Massachusetts
London, England
2016

Printed in the United States of America

Sixth printing

Library of Congress Cataloging-in-Publication Data
Names: Stigler, Stephen M., author.
Title: The seven pillars of statistical wisdom / Stephen M. Stigler.
Description: Cambridge, Massachusetts : Harvard University Press,
    2016. | Includes bibliographical references and index.
Identifiers: LCCN 2015033367 | ISBN 9780674088917 (pbk. : alk. paper)
Subjects: LCSH: Statistics—History. | Mathematical statistics—History.
Classification: LCC QA276.15 S754 2016 | DDC 519.5—dc23
LC record available at http://lccn.loc.gov/2015033367

To my grandchildren,
Ava and Ethan

CONTENTS

THE SEVEN PILLARS OF
STATISTICAL WISDOM

· INTRODUCTION ·

WHAT IS STATISTICS? THIS QUESTION WAS ASKED AS EARLY AS 1838—in reference to the Royal Statistical Society—and it has been asked many times since. The persistence of the question and the variety of answers that have been given over the years are themselves remarkable phenomena. Viewed together, they suggest that the persistent puzzle is due to Statistics not being only a single subject. Statistics has changed dramatically from its earliest days to the present, shifting from a profession that claimed such extreme objectivity that statisticians would only gather data—not analyze them—to a profession that seeks partnership with scientists in all stages of investigation, from planning to analysis. Also, Statistics presents different faces to different sciences: In some applications, we accept the scientific model as derived from mathematical theory; in some, we construct a model that can then take on a status as firm as any Newtonian construction. In some, we are active planners and passive analysts; in others, just the reverse. With so many faces, and the consequent challenges of balance to avoid missteps, it is no wonder that the question, "What is Statistics?" has arisen again and again, whenever a new challenge arrives, be it the economic statistics of the 1830s, the biological questions of the 1930s, or the vaguely defined "big data" questions of the present age.

With all the variety of statistical questions, approaches, and interpretations, is there then no core science of Statistics? If we are fundamentally dedicated to working in so many different sciences, from public policy to validating the discovery of the Higgs boson, and we are sometimes seen as mere service personnel, can we really be seen in any reasonable sense as a unified discipline, even as a science of our own? This is the question I wish to address in this book. I will not try to tell you what Statistics is or is not; I will attempt to formulate seven principles, seven pillars that have supported our field in different ways in the past and promise to do so into the indefinite future. I will try to convince you that each of these was revolutionary when introduced, and each remains a deep and important conceptual advance.

My title is an echo of a 1926 memoir, *Seven Pillars of Wisdom*, by T. E. Lawrence, Lawrence of Arabia.[1] Its relevance comes from Lawrence's own source, the Old Testament's Book of Proverbs 9:1, which reads, "Wisdom hath built her house, she hath hewn out her seven pillars." According to Proverbs, Wisdom's house was constructed to welcome those seeking understanding; my version will have an additional goal: to articulate the central intellectual core of statistical reasoning.

In calling these seven principles the Seven Pillars of Statistical Wisdom, I hasten to emphasize that these are seven *support* pillars—the disciplinary foundation, not the whole edifice, of Statistics. All seven have ancient origins, and the modern discipline has constructed its many-faceted science upon this structure with great ingenuity and with a constant supply of exciting new ideas of splendid promise. But without taking away from that modern work, I hope to articulate a unity at the core of Statistics both across time and between areas of application.

The first pillar I will call Aggregation, although it could just as well be given the nineteenth-century name, "The Combination of Observations," or even reduced to the simplest example, taking a mean. Those simple names are misleading, in that I refer to an idea that is now old but was truly revolutionary in an earlier day—and it still is so today, whenever it reaches into a new area of application. How is it revolutionary? By stipulating that, given a number of observations, you can actually gain information by throwing information away! In taking a simple arithmetic mean, we discard the individuality of the measures, subsuming them to one summary. It may come naturally now in repeated measurements of, say, a star position in astronomy, but in the seventeenth century it might have required ignoring the knowledge that the French observation was made by an observer prone to drink and the Russian observation was made by use of an old instrument, but the English observation was by a good friend who had never let you down. The details of the individual observations had to be, in effect, erased to reveal a better indication than any single observation could on its own.

The earliest clearly documented use of an arithmetic mean was in 1635; other forms of statistical summary have a much longer history, back to Mesopotamia and nearly to the dawn of writing. Of course, the recent important instances of this first pillar are more complicated. The method of least squares and its cousins and descendants are all averages; they are weighted aggregates of data that submerge the identity of individuals, except for designated covariates. And devices like kernel estimates of densities and various modern smoothers are averages, too.

The second pillar is Information, more specifically Information Measurement, and it also has a long and interesting intellectual history. The question of when we have enough evidence to be convinced a medical treatment works goes back to the Greeks. The mathematical study of the rate of information accumulation is much more recent. In the early eighteenth century it was discovered that in many situations the amount of information in a set of data was only proportional to the square root of the number n of observations, not the number n itself. This, too, was revolutionary: imagine trying to convince an astronomer that if he wished to double the accuracy of an investigation, he needed to quadruple the number of observations, or that the second 20 observations were not nearly so informative as the first 20, despite the fact that all were equally accurate? This has come to be called the root-n rule; it required some strong assumptions, and it required modification in many complicated situations. In any event, the idea that information in data could be measured, that accuracy was related to the amount of data in a way that could be precisely articulated in some situations, was clearly established by 1900.

By the name I give to the third pillar, Likelihood, I mean the calibration of inferences with the use of probability. The simplest form for this is in significance testing and the common P-value, but as the name "Likelihood" hints, there is a wealth of associated methods, many related to parametric families or to Fisherian or Bayesian inference. Testing in one form or another goes back a thousand years or more, but some of the earliest tests to use probability were in the early eighteenth century. There were many examples in the 1700s and 1800s, but systematic treatment only came with the twentieth-century work of Ronald A. Fisher and of Jerzy Neyman and Egon S. Pearson, when a full theory of likelihood began serious development. The use of probability to calibrate inference may be most familiar in testing, but it occurs everywhere a number is attached to an inference, be it a confidence interval or a Bayesian posterior probability. Indeed, Thomas Bayes's theorem was published 250 years ago for exactly that purpose.

The name I give the fourth pillar, Intercomparison, is borrowed from an old paper by Francis Galton. It represents what was also once a radical idea and is now commonplace: that statistical comparisons do not need to be made with respect to an exterior standard but can often be made in terms interior to the data themselves. The most commonly encountered examples of intercomparisons are Student's t-tests and the tests of the analysis of variance. In complex designs, the partitioning of variation can be an intricate operation and allow blocking, split plots, and hierarchical designs to be evaluated based entirely upon the data at hand. The idea is quite radical, and the ability to ignore exterior scientific standards in doing a "valid" test can lead to abuse in the wrong hands, as with most powerful tools. The bootstrap can be thought of as a modern version of intercomparison, but with weaker assumptions.

I call the fifth pillar Regression, after Galton's revelation of 1885, explained in terms of the bivariate normal distribution. Galton arrived at this by attempting to devise a mathematical framework for Charles Darwin's theory of natural selection, overcoming what appeared to Galton to be an intrinsic contradiction in the theory: selection required increasing diversity, in contradiction to the appearance of the population stability needed for the definition of species.

The phenomenon of regression can be explained briefly: if you have two measures that are not perfectly correlated and you select on one as extreme from its mean, the other is expected to (in standard deviation units) be less extreme. Tall parents on average produce somewhat shorter children than themselves; tall children on average have somewhat shorter parents than themselves. But much more than a simple paradox is involved: the really novel idea was that the question gave radically different answers depending upon the way it was posed. The work in fact introduced modern multivariate analysis and the tools needed for any theory of inference. Before this apparatus of conditional distributions was introduced, a truly general Bayes's theorem was not feasible. And so this pillar is central to Bayesian, as well as causal, inference.

The sixth pillar is Design, as in "Design of Experiments," but conceived of more broadly, as an ideal that can discipline our thinking in even observational settings. Some elements of design are extremely old. The Old Testament and early

Arabic medicine provide examples. Starting in the late nineteenth century, a new understanding of the topic appeared, as Charles S. Peirce and then Fisher discovered the extraordinary role randomization could play in inference. Recognizing the gains to be had from a combinatorial approach with rigorous randomization, Fisher took the subject to new levels by introducing radical changes in experimentation that contradicted centuries of experimental philosophy and practice. In multifactor field trials, Fisher's designs not only allowed the separation of effects and the estimation of interactions; the very act of randomization made possible valid inferences that did not lean on an assumption of normality or an assumption of homogeneity of material.

I call the seventh and final pillar Residual. You might suspect this is an evasion, "residual" meaning "everything else." But I have a more specific idea in mind. The notion of residual phenomena was common in books on logic from the 1830s on. As one author put it, "Complicated phenomena . . . may be simplified by subducting the effect of known causes, . . . leaving . . . a *residual phenomenon* to be explained. It is by this process . . . that science . . . is chiefly promoted."[2] The idea, then, is classical in outline, but the use in Statistics took on a new form that radically enhances and disciplines the method by incorporating structured families of models and employing the probability calculus and statistical logic to decide among them. The most common appearances in Statistics are our model diagnostics (plotting residuals), but more important is the way we explore high-dimensional spaces by fitting and comparing nested models. Every test for significance of a regression coefficient is an example, as is every exploration of a time series.

At serious risk of oversimplification, I could summarize and rephrase these seven pillars as representing the usefulness of seven basic statistical ideas:

1. The value of targeted reduction or compression of data

2. The diminishing value of an increased amount of data

3. How to put a probability measuring stick to what we do

4. How to use internal variation in the data to help in that

5. How asking questions from different perspectives can lead to revealingly different answers

6. The essential role of the planning of observations

7. How all these ideas can be used in exploring and comparing competing explanations in science

But these plain-vanilla restatements do not convey how revolutionary the ideas have been when first encountered, both in the past and in the present. In all cases they have pushed aside or overturned firmly held mathematical or scientific beliefs, from discarding the individuality of data values, to downweighting new and equally valuable data, to overcoming objections to any use of probability to measure uncertainty outside of games of chance. And how can the variability interior to our data measure the uncertainty about the world that produced it? Galton's multivariate analysis revealed to scientists that their reliance upon rules of proportionality dating from Euclid did not apply to a scientific world in which there was variation in the data—overthrowing three thousand years of mathematical tradition. Fisher's designs were in direct contradiction to what experimental scientists and logicians had believed for centuries; his methods for comparing models were absolutely new to experimental science and required a change of generations for their acceptance.

As evidence of how revolutionary and influential these ideas all were, just consider the strong push-back they continue to attract, which often attacks the very aspects I have been listing as valued features. I refer to:

Complaints about the neglect of individuals, treating people as mere statistics

Implied claims that big data can answer questions on the basis of size alone

Denunciations of significance tests as neglectful of the science in question

Criticisms of regression analyses as neglecting important aspects of the problem

These questions are problematic in that the accusations may even be correct and on target in the motivating case, but they are frequently aimed at the method, not the way it is used in the case in point. Edwin B. Wilson made a nice comment on this in 1927. He wrote, "It is largely because of lack of knowledge of what statistics is that the person untrained in it trusts himself with a tool quite as dangerous as any he may pick out from the whole armamentarium of scientific methodology."[3]

The seven pillars I will describe and whose history I will sketch are fine tools that require wise and well-trained hands for effective use. These ideas are not part of Mathematics, nor are they part of Computer Science. They are centrally of Statistics, and I must now confess that while I began by explicitly denying that my goal was to explain what Statistics is, I may by the end of the book have accomplished that goal nonetheless.

I return briefly to one loose end: What exactly does the passage in Proverbs 9:1 mean? It is an odd statement: "Wisdom hath built her house, she hath hewn out her seven pillars." Why would a house require seven pillars, a seemingly unknown structure in both ancient and modern times? Recent research has shown, I think convincingly, that scholars, including those responsible for the Geneva and King James translations of the Bible, were uninformed on early Sumerian mythology and mistranslated the passage in question in the 1500s. The reference was not to a building structure at all; instead it was to the seven great kingdoms of Mesopotamia before the flood, seven kingdoms in seven cities founded on principles formulated by seven wise men who advised the kings. Wisdom's house was based upon the principles of these seven sages. A more recent scholar has offered this alternative translation: "Wisdom has built her house, The seven have set its foundations."[4]

Just so, the seven pillars I offer are the fruit of efforts by many more than seven sages, including some whose names are lost to history, and we will meet a good selection of them in these pages.

CHAPTER 1

· AGGREGATION ·

From Tables and Means to Least Squares

THE FIRST PILLAR, AGGREGATION, IS NOT ONLY THE OLDEST; it is also the most radical. In the nineteenth century it was referred to as the "combination of observations." That phrase was meant to convey the idea that there was a gain in information to be had, beyond what the individual values in a data set tell us, by combining them into a statistical summary. In Statistics, a summary can be more than a collection of parts. The sample mean is the example that received the earliest technical focus, but the concept includes other summary presentations, such as weighted means and even the method of least squares, which is at bottom a weighted or adjusted average, adjusting for some of the other characteristics of individual data values.

The taking of a mean of any sort is a rather radical step in an analysis. In doing this, the statistician is discarding information in the data; the individuality of each observation is lost: the order in which the measurements were taken and the differing circumstances in which they were made, including the identity of the observer. In 1874 there was a much-anticipated transit of Venus across the face of the sun, the first since 1769, and many nations sent expeditions to places thought to be favorable for the viewing. Knowing the exact time from the beginning to the end of the transit across the sun could help to accurately determine the dimensions of the solar system. Were numbers reported from different cities really so alike that they could be meaningfully averaged? They were made with different equipment by observers of different skills at the slightly different times the transit occurred at different locations. For that matter, are successive observations of a star position made by a single observer, acutely aware of every tremble and hiccup and distraction, sufficiently alike to be averaged? In ancient and even modern times, too much familiarity with the circumstances of each observation could undermine intentions to combine them. The strong temptation is, and has always been, to select one observation thought to be the best, rather than to corrupt it by averaging with others of suspected lesser value.

Even after taking means had become commonplace, the thought that discarding information can increase information has not always been an easy sell. When in the 1860s William Stanley Jevons proposed measuring changes in price level by an index number that was essentially an average of the percent changes in different commodities, critics considered it absurd to average data on pig iron and pepper. And once the discourse shifted to individual commodities, those investigators with detailed historical knowledge were tempted to think they could "explain" every movement, every fluctuation, with some story of why that particular event had gone the way it did. Jevons's condemnation of this reasoning in 1869 was forceful: "Were a complete explanation of each fluctuation thus necessary, not only would all inquiry into this subject be hopeless, but the whole of the statistical and social sciences, so far as they depend upon numerical facts, would have to be abandoned."[1] It was not that the stories told about the data were false; it was that they (and the individual peculiarities in the separate observations) had to be pushed into the background. If general tendencies were to be revealed, the observations must be taken as a set; they must be combined.

Jorge Luis Borges understood this. In a fantasy short story published in 1942, "Funes the Memorious," he described a man, Ireneo Funes, who found after an accident that he could remember absolutely everything. He could reconstruct every day in the smallest detail, and he could even later reconstruct the reconstruction, but he was incapable of understanding. Borges wrote, "To think is to forget details, generalize, make abstractions. In the teeming world of Funes there were only details."[2] Aggregation can yield great gains above the individual components. Funes was big data without Statistics.

When was the arithmetic mean first used to summarize a data set, and when was this practice widely adopted? These are two very different questions. The first may be impossible to answer, for reasons I will discuss later; the answer to the second seems to be sometime in the seventeenth century, but being more precise about the date also seems intrinsically difficult. To better understand the measurement and reporting issues involved, let us look at an interesting example, one that includes what may be the earliest published use of the phrase "arithmetical mean" in this context.

Variations of the Needle

By the year 1500, the magnetic compass or "needle" was firmly established as a basic tool of increasingly adventurous mariners. The needle could give a reading on magnetic north in any place, in any weather. It was already well known a century earlier that magnetic north and true north differed, and by 1500 it was also well known that the difference between true and magnetic north varied from place to place, often by considerable amounts—10° or more to the east or to the west. It was at that time believed this was due to the lack of magnetic attraction by the sea and the consequent bias in the needle toward landmasses and away from seas. The correction needed to find true north from a compass was called the variation of the needle. Some navigational maps of that period would mark the known size of this correction at key locations, such as in straits of passage and near land-marks visible from sea, and mariners had confidence in these recorded deviations. William Gilbert, in his classic book on terrestrial magnetism published in 1600, *De Magnete*, reported that the constancy of the variation at each location could be counted on as long as the earth was stable: "As the needle ever inclined toward east or toward west, so even now does the arc of variation continue to be the same in whatever place or region, be it sea or continent; so too, will it be forever-more unchanging, save there should be a great break-up of a continent and annihilation of countries, as the region At-lantis, whereof Plato and ancient writers tell."[3]

Alas, the mariners and Gilbert's confidence was misplaced. In 1635, Henry Gellibrand compared a series of determi-nations of the variation of the needle at the same London location at times separated by more than fifty years, and he found that the variation had changed by a considerable amount.[4] The correction needed to get true north had been

11° east in 1580, but by 1634 this had diminished to about 4° east. These early measurements were each based upon several observations, and a closer look at them shows how the observers were separately and together groping toward a use of the arithmetic mean, but never clearly getting there.

The best recorded instance of these early determinations of the variation of the needle was published by William Borough in 1581 in a tract entitled *A Discours of the Variation of the Cumpas, or Magneticall Needle.*[5] In chapter 3 he described a method to determine a value for the variation without detailed preliminary knowledge of the direction of true north at one's location, and he illustrated its use at Limehouse, in the Docklands at the East End of London, not far from the Greenwich Meridian. He suggested making careful observations of the sun's elevation with an astrolabe (essentially a brass circle marked with a scale in degrees, to be suspended vertically while the sun was observed with a finder scope and its elevation noted). Every time the sun reached a new degree of elevation in ascent before noon and in descent after noon, Borough would take a reading of the deviation of the sun from magnetic north by noting and recording the direction of a shadow of a wire on his magnetic compass's face. The sun's maximum elevation should be achieved when the sun is at the meridian, when it was at true north (see Figure 1.1).

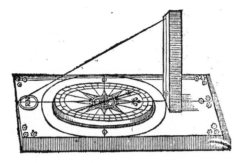

1.1 The compass used by Borough. The vertical column is at the north end of the compass, which is marked by a fleur-de-lis. The initials R. N. are those of Robert Norman, to whose book Borough's tract was appended. The "points" of the compass he refers to in the text are not the eight points as shown but the division lines dividing each of their gaps into four parts and thus dividing the circle into 32 parts, each 11° 15' in size. *(Norman 1581)*

Borough would consider each pair of recorded compass observations made at the same elevation of the sun, one in the morning (Fornoone in Figure 1.2, also designated AM below) and one in the afternoon (Afternoone, designated PM below). On the one hand, if true and magnetic north agree at Limehouse, that common value should be (nearly) the midpoint between the two measurements, since the sun travels a symmetrical arc with the maximum at the meridian ("high noon"). On the other hand, if magnetic north is 10° east of true north, then the morning shadow should be 10° farther west and the afternoon shadow likewise. In either case the average of the two should then give the

¶ *Jn Limehouſe the ſixteenth of*
October. Anno,1 5 8 0.

Fornoone.			Afternoone.			
Elevation of the Sunne.	*Variation of the ſhadow from the North of the Needle to the Weſtwards.*		*Elevation of the Sunne.*	*Variation of the ſhadow from the North of the Needle to the Eaſtwards.*		*Variation of the Needle from the Pole or Axis.*
Deg.	Degr.	Min.	Deg.	D.	M.	D. M.
17	52	35	17	30	0	11 17½
18	50	8	18	27	45	11 11¼
19	47	30	19	24	30	11 30
20	45	0	20	22	15	11 22½
21	42	15	21	19	30	11 22¼
22	38	0	22	15	30	11 15
23	34	40	23	12	0	11 20
24	29	35	24	7	0	11 17
25	22	20	25	Fró N.to w. 0,8′		11 14

1.2 Borough's 1580 data for the variation of the needle at Limehouse, near London. *(Norman 1581)*

variation of the needle. Borough's table of data for October 16, 1580, is presented in Figure 1.2.

He had data for nine pairs, taken at elevations from 17° to 25° with the morning variations (given in westward degrees) and afternoon variations (given in eastward degrees, so opposite in sign to the morning, except for the 25° afternoon measure, which was slightly westward). Because of the different signs in the morning and afternoon, the variations in the right-hand column are found as the difference of the variations divided by 2. For the pair taken at sun's elevation 23° we have

$$(AM + PM)/2 = (34° 40' + (-12° 0'))/2$$
$$= (34° 40' - 12° 0')/2$$
$$= (22° 40')/2 = 11° 20'.$$

The nine determinations are in quite good agreement, but they are not identical. How could Borough go about determining a single number to report? In a pre-statistical era, the need to report data was clear, but as there was no agreed upon set of summary methods, there was no need to describe summary methods—indeed, there was no precedent to follow. Borough's answer is simple: referring to the right hand column, he writes, "conferring them all together, I do finde the true variation of the Needle or Cumpas at Lymehouse to be about 11 d. $\frac{1}{4}$, or 11 d. $\frac{1}{3}$, whiche is a poinct of the Cumpas just or a little more." His value of 11 d. 15 m. (11° 15') does not correspond to any modern summary measure—it is smaller than the mean, median, midrange, and mode. It agrees with the value for 22° elevation, and could have been so chosen—but then why also give 11 d. 20 m., the figure for 23° elevation? Or perhaps he rounded to agreement with "one point of the compass," that is, the 11 d. 15 m. distance between each of the 32 points of the compass? Regardless, it is clear Borough did not feel the necessity for a formal compromise. He could take a mean of two values from morning and afternoon at the same elevation, but that was a clever way of using contrasting observations to arrive at a result, not a combination of essentially equivalent observations. That average was a "before minus after" contrast.

In 1634, more than half a century later, the Gresham College professor of astronomy Gellibrand revisited the problem (see Figure 1.3). Twelve years earlier, Gellibrand's predecessor at Gresham, Edmund Gunter, had repeated Borough's experiment at Limehouse and made eight determinations of the variation in the needle, with results ranging around 6°, quite different from Borough's 11¼. Gunter was an excellent observer, but he lacked the imagination to see this as a discovery, attributing the discrepancy between his results and Borough's to errors by Borough. Gellibrand had too high an opinion of Borough to endorse that view, writing with regret that "this great discrepance [has] moved some of us to be overhasty in casting an aspersion of error on Mr Burrows observations (though since upon noe just grounds)."[6] Gellibrand did try adjusting Borough's figures for the parallax of the sun using a method he attributed to Tycho Brahe that had not been available to Borough, but the effect was negligible (for example, Borough's value for 20 d. elevation, namely 11 d. 22½ m., became about 11 d. 32½ m.). Gellibrand then set about making his own observations with fancy new equipment (including a six-foot quadrant instead of

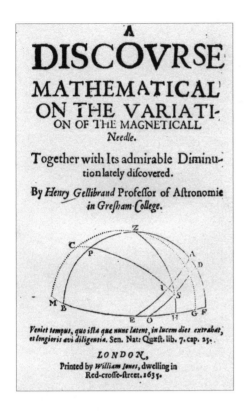

1.3 The title page of Gellibrand's tract. (Gellibrand 1635)

an astrolabe) at Deptford, just south of the Thames from (and at the same longitude as) Limehouse.

On June 12, 1634, using methods based upon Brahe's tables, Gellibrand made eleven separate determinations of the variation of the needle: five before noon and six after noon (see Figure 1.4). The largest was $4° 12'$; the smallest was $3° 55'$. He summarized as follows:

Obfervations made at Diepford An. 1634 Iunij 12 beforeNoone

Alt: ⊙ vera	Azim: Mag	Azim. ⊙	variatio
Gr. Min.	Gr. ℳ.	Gr. ℳ,	Gr. ℳ.
44, 45.	106, 0	110 6	4. 6
46, 30,	109, 0	113 10	4, 10
48, 31,	113, 0	117 1	4. 1
50, 54,	118 0	122, 3	4. 3
54, 24,	127 0	130 55	3 55

After Noone the fame day.

Alt. ⊙ vera	Azi. Mag	Azim. ⊙	Variation
Gr. Min.	Gr: ℳ.	G, ℳn.	Gr, Min
44 37	114: 0	109. 53.	4: 7
40 48	108: 0	103, 50	4: 10
38 46	105. 0	100, 48	4. 12
36 43	102, 0	97. 56	4. 4
34 32	99, 0	95, 0	4: 0
32 10	96: 0	91. 55	4: 5

Thefe Concordant Obfervations can not produce a variation greater then 4 gr. 12 min. nor leffe then 3 gr. 55 min. the Arithmeticall meane limiting it to 4 gr. and about 4 minutes.

1.4   *(above and below)* Gellibrand's data and the appearance of "Arithmeticall meane." *(Gellibrand 1635)*

These Concordant Observations can not produce a variation greater than 4 gr. 12 min. nor lesse than 3 gr. 55 min. the Arithmeticall meane limiting it to 4 gr. and about 4 minutes.[7] ["gr." here simply refers to "degree", the unit of the "graduated" scale at that time. In the 1790s a French revolutionary scale would use "gr." for "grad" to mean $\frac{1}{100}$ of a right angle.]

The "meane" Gellibrand reports, then, is not the arithmetic mean of all eleven; that would be $4° 5'$. Instead he gives the mean of the largest and smallest, what later statisticians would call a midrange. As such it is not remarkable. While it is an arithmetic mean of two observations, there is scarcely any other way of effecting a compromise between two values. There were in fact several earlier astronomers who had done this or something similar when confronted with two values and in need of a single value—certainly Brahe and Johannes Kepler in the early 1600s, and possibly al-Biruni ca. 1000 CE. What was new with Gellibrand's work was the terminology—he gives a name to the method used. The name had been known to the ancients, but, as far as is now known, none of them had felt it useful or necessary to actually use the name in their written work.

A later sign that the statistical analysis of observations had really entered into a new phase was a short note in the *Transactions of the Royal Society* in 1668, also on the variation of the needle. There, the editor, Henry Oldenburg, printed an extract from a letter from a "D. B." that gave five values for the variation at a location near Bristol (see Figure 1.5).

An Extract
Of a Letter, written by D. B. to the Publisher, concerning the pre-
sent Declination of the Magnetick Needle, and the Tydes,
May 23. 1668.

SIr, I here present you with a Scheme of the *Magnetical Vari-*
ations, as it was sent me by Capt. *Samuel Sturmy*, an experi-
enced Seaman, and a Commander of a Merchant Ship for many
years; who (as he assures
me) took the Observati-
ons himself in the presence
of Mr. *Staynred*, an antient
Mathematician, & others,
in *Rownham*-Meadowes by
the water-side, in some
such approach, I think, to
Bristol, as *Lime-house* or
the Fields adjoyning are to
London. This (as the
Table shews) was taken
June 13. 1666. They ob-

Observed June 13. 1668.			
Sun's-Observ'd Altitude.	Magne-tical Azimuth.	Suns true Azimuth.	Variat. Wester-ly.
Gr. M.	Gr. M.	Gr. M.	G. M.
44 20	72 00	70 38	1 22
39 30	80 00	78 24	1 36
31 50	90 00	88 26	1 34
27 42	95 00	93 36	1 24
23 20	103 00	101 23	1 23

served again in the same day of the next year, *viz.* *June* 13.
1667; and then they found the Variation increas'd about 6. minutes *Westerly.*

1.5 The opening passage
of D. B.'s letter.
(D. B. 1668)

D. B. reports Captain Sturmy's summary: "In taking this
Table he notes the greatest distance or difference to be 14
minutes; and so taking the *mean* for the true *Variation*, he
concludes it *then* and *there* to be just 1 deg. 27 min. viz. June
13 1666."[8] While the true mean is 1° 27.8' and Captain
Sturmy (or the mathematician Staynred) rounded down, it
is in any event clear that the arithmetic mean had arrived
by the last third of that century and been officially recognized as a method for combining observations. The date
of birth may never be known, but the fact of birth seems
undeniable.

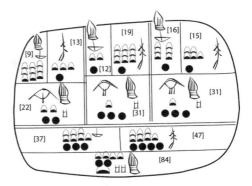

1.6 A reconstruction of a ca. 3000 BCE Sumerian tablet, with modern numbers added. *(Reconstruction by Robert K. Englund; from Englund 1998, 63)*

Aggregation in Antiquity

Statistical summaries have a history as long as that of writing. Figure 1.6 is a reconstruction of a Sumerian clay tablet dating from about 3000 BCE—nearly the dawn of writing—shown to me by a colleague, Chris Woods, at the Oriental Institute at the University of Chicago.

The tablet presents what amounts to a 2×3 contingency table, showing counts of two types of a commodity, possibly the yields of two crops over three years (with modern numbers added).[9] The top row shows the six cells, with commodity symbols above the respective counts. In the second row are the year or column totals, in the third row are the two crop row totals, and the grand total is at the

	Year 1	Year 2	Year 3	Total
Crop A	9	12	16	37
Crop B	13	19	15	47
Total	22	31	31	84

bottom. We would today arrange the numbers differently, as the table shows.

The statistical analysis has not survived, but we can be certain it did not include a chi-square test. What we can say is that the tablet shows a high level of statistical intelligence for its time, but it does not move far away from the individual data values: not only does the body of the table show all the crop-per-year counts, the back side of the tablet gives the raw data upon which these counts were based—the individual producer numbers. Even five thousand years ago some people felt it useful to publish the raw data!

But when did the scientific analysis of statistical data begin? When did the use of the arithmetic mean become a formal part of a statistical analysis? Was it really not much before the seventeenth century? Why was the mean not used to combine observations in some earlier era—in astronomy, surveying, or economics? The mathematics of a mean was certainly known in antiquity. The Pythagoreans knew already in 280 BCE of three kinds of means: the arithmetic, the geometric, and the harmonic. And by 1000 CE the philosopher Boethius had raised that number to at least ten, including the Pythagorean three. To be sure, these means were deployed in philosophical senses, in discussing proportions between line segments, and in music, not as data summaries.

Surely we might expect the Greeks or the Romans or the Egyptians to have taken means of data in day-to-day life more than two millennia ago. Or, if they did not, surely the mean may be found in the superb astronomical studies of Arabic science a thousand years ago. But diligent and far-ranging searches for even one well-documented example from those sources have come up empty.

The most determined seeker of an early use of the mean was the indefatigable researcher Churchill Eisenhart, who spent most of his professional career at the National Bureau of Standards. Over several decades he pursued historical uses of the mean, and he summarized his researches in his presidential address to the American Statistical Association in 1971.[10] His enthusiasm carried the address to nearly two hours, but for all that effort, the earliest documented uses of the mean he found were those I have already mentioned by D. B. and Gellibrand. Eisenhart found that Hipparchus (ca. 150 BCE) and Ptolemy (ca. 150 CE) were silent on their statistical methods; al-Biruni (ca. 1000 CE) gave nothing closer to a mean than using the number produced by splitting the difference between a minimum and a maximum. The mean occurred in practical geometry in India quite early; Brahmagupta, in a tract on mensuration written in 628 CE, suggested approximating the volume of irregular excavations using that of the rectangular solid with the excavation's mean dimensions.[11]

Over all these years, the historical record shows that data of many types were collected. In some cases, inevitably, summaries would be needed; if the mean was not used, what did people do in order to summarize, to settle on a single figure to report? Perhaps we can get a better idea of how this question was seen in pre-statistical times by looking at a few examples where something similar to a mean was employed.

One story told by Thucydides involved siege ladders and dates to 428 BCE:

> Ladders were made to match the height of the enemy's wall, which they measured by the layers of bricks, the side turned towards them not being thoroughly whitewashed. These were counted by many persons at once; and though some might miss the right calculation, most would hit upon it, particularly as they counted over and over again, and were no great way from the wall, but could see it easily enough for their purpose. The length required for the ladders was

thus obtained, being calculated from the breadth of the brick.[12]

Thucydides described the use of what we can recognize as the mode—the most frequently reported value. With the expected lack of independence among counts, the mode is not particularly accurate, but if the reports were tightly clustered it was likely as good as any other summary. Thucydides did not give the data.

Another, much later example dates from the early 1500s and is reported by Jacob Köbel in a finely illustrated book on surveying. As Köbel tells us, the basic unit of land measure in those times was the rod, defined as sixteen feet long. And in those days a foot meant a real foot, but whose foot? Surely not the king's foot, or each change of monarch would require a renegotiation of land contracts. The solution Köbel reports was simple and elegant: sixteen representative citizens (all male in those days) would be recruited after a church service and asked to stand in a line, toe to heel, and the sixteen-foot rod would be the length of that line. Köbel's picture, etched by himself, is a masterpiece of explanatory art (see Figure 1.7).[13]

It was truly a community rod! And, after the rod was determined, it was subdivided into sixteen equal sections, each representing the measure of a single foot, taken from the

1.7 Köbel's depiction of the determination of the lawful rod. *(Köbel 1522)*

communal rod. Functionally, this was the arithmetic mean of the sixteen individual feet, but nowhere was the mean mentioned.

These two examples, separated by nearly two millennia, involve a common problem: how to summarize a set of similar, but not identical, measurements. The way the problem was dealt with in each situation reflects the intellectual difficulty involved in combination, one that persists today. In antiquity, and in the Middle Ages, when reaching for a summary of diverse data, people chose an individual example. In Thucydides's story, the individual case chosen was the most popular case, the mode. So it was in other instances: the case selected might be a prominent one—for numeric data it could even be a maximum, a record value. Every society has wished to trumpet their best as representing the whole. Or the selected case could be simply an individual or value picked as "best" for reasons not well articulated. In astronomy, the selection of a "best" value could reflect personal knowledge of the observer or the atmospheric conditions for observation. But whatever was done, it amounted to maintaining the individuality of at least one data value. In Köbel's account the emphasis was on the sixteen individual feet; the people in the picture may have even been recognizable at the time. In any event, the idea that the individuals were collectively determining the rod was the forceful point—their identity was not discarded; it was the key to the legitimacy of the rod, even as the separate foot marks were a real average.

The Average Man

By the 1800s, the mean was widely in use in astronomy and geodesy, and it was in the 1830s that it made its way into society more broadly. The Belgian statistician Adolphe Quetelet was at that time crafting the beginnings of what he would call Social Physics, and, to permit comparisons among population groups, he introduced the Average Man. Originally he considered this as a device for comparing human populations, or a single population over time. The average height of an English population could be compared to that of a French population; the average height at a given age could be followed over time to derive a population growth curve. There was no single Average Man; each group would have its own. And yes, he did focus on men; women were spared this reduction to a single number.[14]

Already in the 1840s a critic was attacking the idea. Antoine Augustin Cournot thought the Average Man would be a physical monstrosity: the likelihood that there would be any real person with the average height, weight, and age of a population was extremely low. Cournot noted that if one averaged the respective sides of a collection of right triangles, the resulting figure would not be a right triangle (unless the triangles were all proportionate to one another).

Another critic was the physician Claude Bernard, who wrote in 1865:

> Another frequent application of mathematics to biology is the use of averages which, in medicine and physiology, leads, so to speak, necessarily to error. . . . If we collect a man's urine during 24 hours and mix all his urine to analyze the average, we get an analysis of a urine that simply does not exist; for

urine, when fasting, is different from urine during digestion. A startling instance of this kind was invented by a physiologist who took urine from a railroad station urinal where people of all nations passed, and who believed he could thus present an analysis of average European urine![15]

Quetelet was undeterred by such criticism, and he insisted that the Average Man could serve as a "typical" specimen of a group, capturing the "type," a group representative for comparative analysis. As such, it has been highly successful and often abused. The Average Man and his descendants amount to a theoretical construction that allows some of the methods of physical science to be employed in social science.

In the 1870s, Francis Galton took the idea of a mean a step further, to nonquantitative data. He put considerable time and energy into the construction of what he called "generic images" based upon composite portraiture, where, by superimposing pictures of several members of a group, he essentially produced a picture of the average man or woman in that group (see Figure 1.8).[16] He found that close facial agreement among sisters and other family members permitted a family type to emerge, and he experimented with other groups, producing composites of medals of Alexander the Great (in hopes of revealing a more realistic picture), and of groups of criminals or of people suffering from the same disease.

SPECIMENS OF COMPOSITE PORTRAITURE

PERSONAL AND FAMILY.

Alexander the Great From 6 Different Medals.

Two Sisters.

From 6 Members of same Family Male & Female.

HEALTH. DISEASE. CRIMINALITY.

23 Cases. Royal Engineers, 12 Officers. 11 Privates

Tubercular Disease

6 Cases

9 Cases

8 Cases

1 Cases

2 Of the many Criminal Types

CONSUMPTION AND OTHER MALADIES

I 20 Cases

II 36 Cases

56 Cases Co-composite of I & II

Consumptive Cases.

100 Cases

50 Cases

Not Consumptive.

1.8 Some of Galton's composite portraits. *(Galton 1883)*

Galton practiced restraint in constructing these pictures, and he was well aware of some of the limitations of these generic portraits. As he explained, "No statistician dreams of combining objects of the same generic group that do not cluster towards a common centre; no more should we at-

1.9 Pumpelly's composite portrait of 12 mathematicians. *(Pumpelly 1885)*

tempt to compose generic portraits out of heterogeneous elements, for if we do so the result is monstrous and meaningless."[17] Some who followed him were not so cautious. An American scientist, Raphael Pumpelly, took photographs of those attending a meeting of the National Academy of Sciences in April 1884; the following year he published the results. As one example, in Figure 1.9, the images of 12 mathematicians (at the time, the term included astronomers and physicists) are superimposed as a composite picture of an average mathematician.[18] Leaving aside the fact that the composite picture might appear as sinister as those of Galton's criminals, we could note that the combination of images of clean-shaven individuals with a few of those with full beards and several more of those with mustaches produces a type that looks more like someone who has been lost in the brush for a week.

Aggregation and the Shape of the Earth

In the mid-1700s, the use of statistical aggregation was expanded to situations where the measurements were made under very different circumstances; indeed, it was forced upon scientists by those circumstances. A prime example of the simplest type is the eighteenth-century study of the shape of the earth. To a good first approximation, the earth is a sphere, but with increases in the precision of navigation and astronomy, questions were bound to appear. Isaac Newton, from dynamical considerations, had suggested the earth was a slightly oblate spheroid (compressed at the poles, expanded at the equator). The French astronomer Domenico Cassini thought it was a prolate spheroid, elongated at the poles. The matter could be settled by comparing measures taken on the ground at different latitudes. At several locations from the equator to the North Pole, a relatively short arc length, A, would be measured, the arc being in a direction perpendicular to the equator, a segment of what was called a meridian quadrant, which would run from the North Pole to the equator. The arc length along the ground would be measured and divided by the difference between latitudes at both ends, so it was given as arc length per degree latitude. Latitude was found by sight, the angle between the polar North Star and the horizontal. By seeing how this 1° arc varied depending on how far from the equator you were, the question could be settled.[19]

The relation between arc length and latitude for spheroids is given by an elliptic integral, but for short distances (and only relatively short distances were practical to measure), a simple equation sufficed. Let A = the length as measured along the ground of a 1° arc. Let L = the latitude of the midpoint of the arc, again as determined by observation of the polestar, $L = 0°$ for the equator, $L = 90°$ for the North Pole. Then, to a good approximation you should have $A = z + y \cdot \sin^2 L$ for each short arc measured:

> If the earth were a perfect sphere, $y = 0$ and all 1° arcs are the same length z.
>
> If the earth were oblate (Newton) then $y > 0$ and the arcs vary from length z at the equator ($\sin^2 0° = 0$) to $z + y$ at the North Pole ($\sin^2 90° = 1$).
>
> If the earth is prolate (Cassini), $y < 0$.

The value y could be thought of as the polar excess (or, if negative, deficiency). The "ellipticity" (a measure of departure from a spherical shape) could be calculated approximately as $e = y/3z$ (a slightly improved approximation $e = y/(3z + 2y)$ was sometimes used).

Data were required. The problem sounds easy: measure any two degrees, perhaps one at the equator, and another near Rome. At that time, lengths were measured in Toise, a pre-metric unit where a Toise is about 6.39 feet. A degree of latitude was about 70 miles in length, too large to be practical, so a shorter distance would be measured and extrapolated. A French expedition in 1736 by Pierre Bouguer took measurements near Quito in what is now Ecuador, where it was feasible to measure relatively long distances near the equator in a north–south direction, and he reported a length $A = 56751$ Toise with $\sin^2(L) = 0$. A measurement near Rome by the Jesuit scholar Roger Joseph Boscovich in 1750 found $A = 56979$ Toise with $\sin^2(L) = .4648$. These gave the two equations

$$56751 = z + y \cdot 0$$

$$56979 = z + y \cdot .4648$$

These equations are easily solved to give $z = 56751$ and $y = 228/.4648 = 490.5$, and $e = 490.5/(3 \cdot 56751) = 1/347$, as they tended to write it at the time.

However, by the time Boscovich wrote reports about the question in the late 1750s, there were not two but five repu-

Locus obser- vationis	Latitu- do		$\frac{1}{2}$ fin. vers.rad.	Hexa- pedæ	Differ. a pri- mo	Differ. com- putata	Error
	o	,	10000				
In America	0	0	0	56751	0	0	0
Ad Prom. B. S.	33	18	2987	57037	286	240	−46
In Italia	42	59	4648	56979	228	372	144
In Gallia	49	23	5762	57074	323	461	138
In Lapponia	66	19	8386	57422	671	671	0

1.10   Boscovich's data for five arc lengths. The columns give (in our notation) for each of the arcs $i = 1$ to 5, the latitude L_i (in degrees), $\sin^2(L_i)$ [$= \frac{1}{2}(1 - \cos(L_i)) = \frac{1}{2}$ versed sine], A_i ("hexapedae," the length in Toise), the difference $A_i - A_1$, the same difference using the solution from arcs 1 and 5, and the difference between these differences. The value for $\sin^2(L)$ for the Cape of Good Hope should be 3014, not 2987. *(Boscovich 1757)*

tably recorded arc lengths: Quito ("In America"), Rome ("In Italia"), Paris ("In Gallia"), Lapland ("In Lapponia"), and all the way south to the Cape of Good Hope at the tip of Africa ("Ad Prom. B. S.").[20] Any two of these would give a solution, and so Boscovich was faced with an embarrassment of data: ten solutions, all different (see Figures 1.10 and 1.11).

Boscovich was faced with a dilemma. The five measured arcs were inconsistent. Should he simply pick one pair and settle for what it told him? Instead, he invented a genuinely novel method of aggregation that led him to a principled solution based upon all five. For him the most suspect element in the data was the arc measurement. These arcs required

Binarium	Differ. in pol., & æqu.	Ellipti-citas		Binarium	Differ. in pol., & æqu.	Ellipti-citas
1, & 5	800	$\frac{1}{213}$		2, & 4	133	$\frac{1}{128}$
2, 5	713	$\frac{1}{239}$		3, 4	853	$\frac{1}{200}$
3, 5	1185	$\frac{1}{144}$		1, 3	491	$\frac{1}{347}$
4, 5	1327	$\frac{1}{128}$		2, 3	−350	$-\frac{1}{486}$
1, 4	549	$\frac{1}{314}$		1, 2	957	$\frac{1}{78}$

1.11 Boscovich's calculations for the 10 pairs of arcs, giving for each the polar excess y and the ellipticity $e = 3y/z$. The ellipticities for (2,4) and (1,2) are misprinted and should be 1/1,282 and 1/178. The figures for (1,4) are in error and should be 560 and 1/304. (Boscovich 1757)

careful measures taken under extremely difficult conditions—from the forests around Paris and Rome, to the tip of Africa, to the frozen tundra of Lapland, to the plains of Ecuador halfway around the world—and it was not possible to repeat the measures as a check. Thinking in terms of the equation $A = z + y \sin^2(L)$, Boscovich reasoned as follows: Each choice of z and y implies a corresponding value for A, and the difference between that value and the observed value could be thought of as an adjustment that needed to be made to the observed A if that measure were to fit the equation. Among all possible z and y, which values implied the smallest total of the absolute values of the adjustments, supposing also that the chosen z and y were consistent with mean of the As and the mean of the Ls? Boscovich gave a clever algorithm to solve for the best values, an early instance of what is now called a linear programming problem. For the five arcs, his method gave the answers $z = 56,751$, $y = 692$, $e = 1/246$.

Over the next half century, a variety of ways were suggested for reconciling inconsistent measures taken under different conditions through some form of aggregation. The most successful was the method of least squares, which is formally a weighted average of observations and had the advantage over other methods of being easily extendable to more complicated situations in order to determine more than two unknowns. It was first published by Adrien-Marie Legendre in 1805, and even though it first appeared in a book explaining how to determine the orbits of comets, the illustrative worked example he gave was for the determination of the earth's ellipticity, using the same measures that were taken to define the length of the meter after the French Revolution.[21] Those data gave the ellipticity as $1/148$, a much larger value, but because of the shorter range of the arcs (only 10° latitude, all within France) and the discrepancy from other values, this was judged less suitable than the earlier measures made over the range from the equator to Lapland, so the final meter was determined using a hybrid derived from the different expeditions.

Aggregation has taken many forms, from simple addition to modern algorithms that are opaque to casual inspection. However, the principle of using summaries in place of full enumeration of individual observations, of trying to gain information by selectively discarding information, has remained the same.

CHAPTER 2

· INFORMATION ·

Its Measurement and Rate of Change

related to the first: If we gain information by combining observations, how is the gain related to the number of observations? How can we measure the value and acquisition of information? This also has a long and interesting intellectual history, going back to the ancient Greeks.

The paradox of the heap was well known to the Greeks: One grain of sand does not make a heap. Consider adding one grain of sand to a pile that is not a heap—surely adding only one grain will not make it into a heap. Yet everyone agrees that somehow sand does accumulate in heaps. The paradox is generally attributed to the fourth-century BCE philosopher Eubulides of Miletus; five centuries later the physician and philosopher Galen posed it as a problem with statistical relevance. Galen presented a debate between an empiricist and a dogmatist.[1]

The dogmatist was an early medical theoretician—he would prescribe treatment using logic: Did the symptoms suggest a lack of heat or a surfeit of heat? Then you should heat or cool the patient accordingly. Did there seem to be a toxic element in the body? Then purge by bleeding or some other method.

The empiricist was a proponent of evidence-based medicine: when in doubt about treatments, look at the record. How many times had bleeding or heating been effective? Had the treatment worked before? Had it failed before? When sufficient favorable evidence had accumulated in favor of a treatment, it could be adopted as a standard; until then, remain skeptical.

The dogmatist countered with the paradox of the heap: Surely one favorable case was not sufficient to draw a general conclusion. And if you were at any stage uncertain, then how could the addition of one more favorable case swing the balance? Would you not then have been convinced by but a single case? But then how could you ever be convinced by the accumulation of evidence? And yet, just as there were undeniable heaps of sand, there were convincing treatment records. Galen's sympathies were with the empiricist, and

surely evidence gathered over medical history was given
due attention, but the problem remained. Granted that more
evidence is better than less, but how much better? For a very
long time, there was no clear answer.

The Trial of the Pyx

As an illustration of the costs that arose from the lack of
an answer, consider the trial of the Pyx.[2] In England in the
twelfth century, there was no single, strong, central authority,
and this posed a challenge for monetary policy. There was
indeed a king, but his authority was counterbalanced by that
of several powerful barons, who would eventually force King
John to cede authority through the Magna Carta in 1215. And
at the same time (or even slightly before—the early history
is a bit murky), there was the commercial need for a generally agreed-upon currency, a coinage that could command
broad respect. The principal source of British coinage was
the London Mint, and, until 1851, when it became the Royal
Mint, it operated independently of the Crown. The king and
the barons would bring gold and silver bullion to the mint
and receive coins in return. To ensure that all was done properly, an indenture from the king was in force, stipulating
the weight and fineness of the coins. And to monitor the
success of the mint in meeting the stipulated standards, the

indentures specified that trials should be held to test the mint's product. These trials were an early instance of monitoring for the maintenance of quality in a production process.

The mint's trials date at least to the late 1200s and possibly to a century earlier. The more detailed descriptions of procedure come from a later time, but there is no reason to believe they changed much before the modern era. Each day of production, a selection of coins would be set aside in a box called the Pyx for later testing. The selection was not rigorously random, but the words employed in some accounts (for example, "impartially" or "taken at hazard") suggest that it was not too far off from being a random sample. At various intervals (for example, every three months in the 1300s), the Pyx would be opened in the presence of judges representing parties with an interest in the accuracy of the coinage. Again a selection would be made, some coins to be tested by assay for the fineness of the gold, the others for a test of their weight. It is the latter test that attracts the most statistical interest.

All parties understood that there would inevitably be some variation in the weight of different coins, and the indenture specified both a target weight (let us denote that by T) and an allowed tolerance it called the "remedy" (denote that by R). If the weight was below $T-R$, the master of the mint would be required to make an appropriate forfeiture. At times that forfeiture would be a cash settlement prorated for all the coinage since the previous trial of the Pyx, but in early trials there was a threat that his hand would be cut off, or worse. Too-heavy coins were also a problem, as they would be culled in circulation and melted to bullion by alert entrepreneurs. But there was no gain in that for the mint, and the main focus of the trial was on light coins.

The coins would be weighed in batches, possibly reflecting a vague understanding that the weighing of individual coins was not only more laborious but also prone to a larger percent weighing error than was batch weighing. If, say, 100 gold coins were weighed in a batch, clearly the target should be 100T. But what would the remedy be? The choice they made was revealing: the remedy in that case would be simply 100R; only if the batch weighed less than 100T − 100R would the mint fail the test. But modern statistical theory tells us this is mistaken; this is much too generous to the mint. It is so low a standard that an alert master of the mint could aim nearly as low, for example by minting to a target of T − 0.5R, or even T − 0.8R, and run virtually no risk of failing the test. If the coins' weights varied independently (with variation among individual coins being statistically unrelated to one another), an appropriate remedy for a batch of 100 would be 10R, not 100R. With statistically independent weights, the variation increases as the square root of the number of coins. Independent variation may be a naive assumption, but the result it gives is surely closer to the truth than 100R. Some data from 1866 suggest that the remedy for a single coin was set at about the two standard deviation mark; meaning that for a batch of 100 coins the remedy was mistakenly set about 20 standard deviations from the target. Some trials involved weighing a thousand or more coins; the resulting test would have been as safe for the mint as any bureaucrat could hope for, as long as they aimed even a little above $T-R$.

In a nineteenth-century British parliamentary investigation, the officers at the Royal Mint were asked if they minted to a low target, and they assured the investigators they would never do such a thing, although they believed that the French did it. Of course, in the early years of the trial of the Pyx, even the best mathematicians were unaware of what is now sometimes referred to as the root-n rule, where n here would be the number of coins. Of course one master of the mint was a better than average mathematician: Isaac Newton. From 1696 to 1727 he was warden and then master of the mint. And on his death in 1727, Newton left a sizable fortune. But evidently his wealth can be attributed to investments, and there is no reason to cast suspicion that he had seen the flaw in the mint's procedures and exploited it for personal gain.

Abraham de Moivre

The first recognition that the variation of sums did not increase proportionately with the number of independent terms added (and that the standard deviation of means did not decrease inversely with the number of terms) dates to the 1700s. This novel insight, that information on accuracy did not accumulate linearly with added data, came in the 1720s to Abraham de Moivre as he began to seek a way to accurately compute binomial probabilities with a large number of trials. In 1733 this would lead to his famous result, which we now call the normal approximation to the binomial, but by 1730 he already noticed that a crucial aspect of the distribution was tied to root-n deviations.[3] If the binomial frequency function is considered as a curve, the inflection points (which could be considered to govern the spread) come at plus and minus $\sqrt{n}/2$.

The implication of this was clear even to de Moivre: In his fifth corollary to the normal approximation, expanded slightly in the first translation from Latin in the 1738 edition of his *Doctrine of Chances*, the standard deviation made its appearance (albeit unnamed) when he noted that the interval between these inflection points would amount to a total probability of 0.682688 (about 28/41) for large n, and the shorter interval between $+$ or $-\sqrt{(2n)}/4$ would be about an

The DOCTRINE *of* CHANCES.   239

COROLLARY 5.

And therefore we may lay this down for a fundamental Maxim, that in high Powers, the Ratio, which the Sum of the Terms included between two Extreams diftant on both fides from the middle Term by an Interval equal to $\frac{1}{2}\sqrt{n}$, bears to the Sum of all the Terms, will be rightly expref'd by the Decimal 0.682688, that is $\frac{28}{41}$ nearly.

Still, it is not to be imagin'd that there is any neceffity that the number n fhould be immenfely great; for fuppofing it not to reach beyond the 900th Power, nay not even beyond the 100th, the Rule here given will be tolerably accurate, which I have had confirmed by Trials.

But it is worth while to obferve, that fuch a fmall part as is $\frac{1}{2}\sqrt{n}$ in refpect to n, and fo much the lefs in refpect to n as n increafes, does very foon give the Probability $\frac{28}{41}$ or the Odds of 28 to 13; from whence we may naturally be led to enquire, what are the Bounds within which the proportion of Equality is contained; I anfwer, that thefe Bounds will be fet at fuch a diftance from the middle Term, as will be expreffed by $\frac{1}{4}\sqrt{2n}$ very near; fo in the cafe above mentioned, wherein n was fuppofed $= 3600$, $\frac{1}{4}\sqrt{2n}$ will be about 21.2 nearly, which in refpect to 3600, is not above $\frac{1}{169}$-th part: fo that it is an equal Chance nearly, or rather fomething more, that in 3600 Experiments, in each of which an Event may as well happen as fail, the Excefs of the happenings or failings above 1800 times will be no more than about 21.

2.1 De Moivre's fifth corollary, from the second edition of *The Doctrine of Chances.* The final paragraph is an addition to the earlier text in a privately circulated Latin version in 1733. *(De Moivre 1738)*

even bet (see Figures 2.1 and 2.2).[4] No matter what standard of certainty was adopted, whether 68% or 50% (or 90% or 95%, or, as with the recent discovery of the Higgs boson, about 99.9999998%), the estimated accuracy varied as the square root of the number of trials.

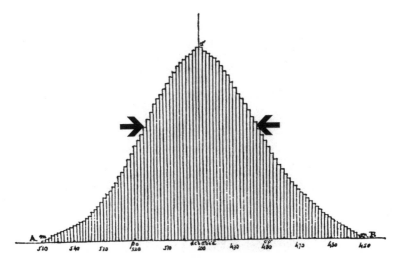

2.2 A picture showing de Moivre's two inflection points (as discussed in 1730) superimposed on a plot of a symmetric binomial with $n=999$, drawn by Quetelet in the 1840s.

In 1810, Pierre Simon Laplace proved a more general form of de Moivre's result, now called the Central Limit Theorem.[5] Where de Moivre had deduced that the number of successes in n binomial trials would vary approximately like a normal curve, Laplace came to the same conclusion in regard to the total or mean of observations (such as the weights of a sample of coins), where the individual observations (or errors in observations) followed pretty much any distribution. The proof was not fully rigorous, and by 1824 Siméon Denis

SUPPLÉMENT AU MÉMOIRE

Sur les approximations des formules qui sont fonctions de très-grands nombres.

Par M. Laplace.

J'ai fait voir dans l'article VI de ce Mémoire, que si l'on suppose dans chaque observation, les erreurs positives et négatives également faciles; la probabilité que l'erreur moyenne d'un nombre n d'observations sera comprise dans les limites $\pm \frac{rh}{n}$, est égale à

$$\frac{2}{\sqrt{\pi}} \cdot \sqrt{\frac{k}{2k'}} \cdot \int dr. \, c^{-\frac{k}{2k'} \cdot r^2}$$

h est l'intervalle dans lequel les erreurs de chaque observation peuvent s'étendre. Si l'on désigne ensuite par $\varphi\left(\frac{x}{h}\right)$ la probabilité de l'erreur $\pm x$, k est l'intégrale $\int dx. \, \varphi\left(\frac{x}{h}\right)$ étendue depuis $x = -\frac{1}{2}h$, jusqu'à $x = \frac{1}{2}h$; k' est l'intégrale $\int \frac{x^2}{h^2} \cdot dx. \, \varphi\left(\frac{x}{h}\right)$, prise dans le même intervalle : π est la demi-circonférence dont le rayon est l'unité, et c est le nombre dont le logarithme hyperbolique est l'unité.

Supposons maintenant qu'un même élément soit donné par n observations d'une première espèce, dans laquelle

2.3 Laplace's first clear statement of the Central Limit Theorem. Here he used c where we now use e, k'/k would be the variance, and the integral was meant to equal the probability that a mean error would not exceed the given limits (circled here). But the n in the denominator should be $\sqrt{n}$. *(Laplace 1810)*

Poisson noticed an exceptional case—what we now call the Cauchy distribution—but, for a wide variety of situations, the result held up and recognition of the phenomenon was soon widespread among mathematical scientists.

Ironically, there was a typo in the very first publication of Laplace's result—it had n instead of root-n (see Figure 2.3). But that was remedied when the book version appeared two years later.

The implications of the root-n rule were striking: if you wished to double the accuracy of an investigation, it was insufficient to double the effort; you must increase the effort fourfold. Learning more was much more expensive than generally believed. Jacob Bernoulli had ceased work on his great book *Ars Conjectandi* with the discovery that it would apparently take 26,000 trials to achieve what he considered an acceptable level of accuracy; the root-n rule was unknown then, and he could not have known that the level of accuracy he sought was unachievable in practice.[6] With time, statisticians learned they had to settle for less and adjusted their expectations accordingly, all while continuing to seek a better understanding of the accumulation of errors or variation. This was in direct contrast to long mathematical practice: in a sequence of mathematical operations, mathematicians would keep track of the maximum error that could have arisen at each step, a quantity that grew as the series grew, while statisticians would allow for a likely compensation of errors, which would in relative terms shrink as the series grew.

Refinements, Extensions, and Paradoxes

By the middle of the nineteenth century, there were refinements of the rule. George Airy, a British astronomer, published a small textbook, *On the Algebraical and Numerical Theory of Errors of Observations and the Combination of Observations,* in 1861 that included a section on "entangled observations," entangled in the sense that the several observations had components in common and thus were, as we would say now, correlated.[7] Airy did show the effect of this relationship on the variances of derived estimates, a step toward understanding the effect of correlation on the amount of information in data.

Charles S. Peirce, an American philosopher and polymath, went a step further when, in 1879, he published a short note on what he called "The Theory of the Economy of Research." Peirce described his goal as follows: "The doctrine of economy, in general, treats of the relations between utility and cost. That branch of it which relates to research considers the relations between the utility and the cost of diminishing the probable error of our knowledge. Its main problem is, how, with a given expenditure of money, time, and energy, to obtain the most valuable addition to our knowledge."[8]

Peirce posed this as a problem in utility theory: Considering two experiments with different standard deviations of the mixed type that Airy had considered (essentially variance components models), both of which furnished vital information, how could you optimize your effort? In the specific case of a reversible pendulum experiment to measure the force of gravity, how should you allocate time between experiments with the heavy end up and those with the heavy end down? Here was an optimization problem, where the criterion being optimized was explicitly a measure of gain in information with correlated observations. He found the experiment should be performed for an equal period of time in each position, and, further, that the duration of the experiment should be "proportional to the distance of the center of mass from the point of support." He closed his note with this admonition: "It is to be remarked that the theory here given rests on the supposition that the object of the investigation is the ascertainment of truth. When an investigation is made for the purpose of attaining personal distinction, the economics of the problem are entirely different. But that seems to be well enough understood by those engaged in that sort of investigation."[9] Presumably the targets of this sarcastic comment recognized themselves in the description.

In any event, the idea that information in data could be measured, that accuracy was related to the amount of data in a way that could be made precise in some situations, was clearly established by 1900. It should not be assumed that there were no challenges. It would be expected that many people continued to believe that the second 20 observations were at least as valuable as the first 20. But an interesting claim from some distinguished sources that was even more extreme went in the opposite direction. The claim stated that in some situations it was better, when you have two observations, to discard one than to average the two! And, even worse, the argument was correct.

In an article published in the Princeton Review in 1878, the Oxford logician John Venn imagined this situation: A ship's captain plans to capture an enemy fort, and he sends two spies to infiltrate and report back on the caliber of cannon in the fort, so that he may have ready the appropriate size of cannonball to defend the fort from recapture. One spy reports the caliber is 8 inches; the other reports 9 inches. Should the captain arrive with 8.5-inch cannonballs? Of course not; they would be unserviceable in either case. Better to toss a coin to settle on one of the two sizes than to average and be sure of failure.[10]

The problem is that the standard analysis, the analysis behind every other example in this chapter, tacitly assumes that the appropriate measure of accuracy is root mean squared error, or its usual surrogate, the standard deviation of the estimate. If the observations are normally distributed about the target value, then all reasonable measures of accuracy will agree, and agree with the root mean square error. But Venn's example is not of that type. With Venn, there is a high premium on being very close, with no increase in penalty for any estimates that err beyond that close interval. For Venn the appropriate measure would be the probability the estimate was within a small number ε of the target. For Venn the choice of estimate E to estimate the caliber C would maximize $\text{Prob}\{|E-C| \le \varepsilon\}$. Francis Edgeworth agreed, and in a short note in 1883 showed that the solution "toss away one observation chosen at random" was superior to taking the mean in some cases other than Venn's discrete example. Edgeworth gave a class of examples for the error distribution that exhibited the paradox and even included some with continuous unimodal densities with all moments finite.[11] They were more peaked at the mode than is usual, but not absurdly so. The measurement of information clearly required attention to the goal of the investigation.

In the twentieth century other less fanciful cases where the root-n rule fails have received attention. One category is time series models, where serial correlation can fool a data analyst by producing apparent patterns that seem compelling, unless you realize that the serial correlation has reduced the effective sample size to well below the number of points plotted. This has led to mistaken discoveries of periodicity, as even non-periodic mechanisms can give the appearance of periodicity when studied for a limited number of cycles. In the 1980s two distinguished geophysicists found what they believed was evidence of a 26-million-year cycle in the extinction rates of small marine life on earth. If real, it would signal some extraterrestrial cause, and one hypothesis was that our sun had a twin that was currently out of sight but rains radiation on us every 26 million years.

That set off a frenzy of interest, including a cover of *Time*, and eager scientists went looking for other statistical evidence. "Seek and ye shall find" works well in unguided data analysis, and papers were written claiming a similar period of reversals of the earth's magnetic field, and other dubious periodicities. In the end it turned out that the first signal found for extinction rates was indeed slightly periodic, but it was due to an artifact of the data, not a passing death star. The data were keyed to a geological dating of epochs over the past 250 million years. For half of that period, the epochs were well determined but their times were not. The authors of the timeline had to divide a 125-million-year stretch of time into 20 subintervals, with an average of 6.25 million years each. But expressing the end points fractionally would exaggerate the accuracy, and so they broke up the period into segments of lengths 6, 6, 6, 7, 6, 6, 6, 7, 6, 6, 6, 7, 6, 6, 6, 7, 6, 6, 6, 7. That artificial periodicity then worked its way though the analysis to give the appearance of the cycle that had set the frenzy off in the first place.[12]

The paradox of the accumulation of information, namely, that the last 10 measurements are worth less than the first 10, even though all measurements are equivalently accurate, is heightened by the different (and to a degree misleading) uses of the term *information* in Statistics and in science. One case in point is the term *Fisher information* in statistical theory. In its simplest form in parametric estimation problems, the Fisher information $I(\theta)$ is the expected value of the square of the score function,[13] defined to be the derivative of the logarithm of the probability density function of the data; that is, $I(\theta) = E[d \log f(X_1, X_2, \ldots X_n; \theta)/d\theta]^2$. (Intuitively, if the probability of the data changes rapidly with θ, then this derivative will tend to be large, and the more sensitive this probability is to θ, the more informative the data will likely be.) This is a marvelous statistical construct; its reciprocal gives under broadly applicable conditions the variance of the best estimate one could hope to find, and thus it sets a gold standard for what we can attain through aggregation in such situations. But from the point of view of assessing the accumulation of information, it is misleading, in that it is expressed in the wrong units—it is on a squared-units scale. It is an additive measure and would say that equally long segments of data are equally informative. Fisher information is consistent with the root-n rule; one need only take its square root.

Another additive measure of information is that introduced by Claude Shannon in the 1940s. Shannon was concerned with a quite different question, namely, problems of coding and signal processing, in which he took it as axiomatic that the measure was additive since the signal to be transmitted was without limit and coded equally informatively throughout the transmission. In the problems of natural and human science the statistician considers, additivity can only hold on scales that are not linear in the size of the data set.

The statistical assessment of the accumulation of information can be quite a complex task, but with care and attention to correlations and scientific objectives, the measurement of information in data—the comparative information in different data sets and the rate of increase in information with an increase in data—has become a pillar of Statistics.

CHAPTER 3

· LIKELIHOOD ·

Calibration on a Probability Scale

A MEASUREMENT WITH NO CONTEXT IS JUST A NUMBER, A meaningless number. The context gives the scale, helps calibrate, and permits comparisons. Of course, we see context-free numbers every day, for example in tabloids where a common feature is called "By the Numbers," followed by a few examples intended to shock or amuse the reader. Even the august magazine *Science* has adopted that device, and one entry (August 5, 2011) read:

42,000 Number of children in the world that die of celiac disease, according to a PLoS ONE study.

On the face of it, this is a troubling statistic. Or is it? Over what period of time—a week, a year, a decade? And is this a large figure or a small one; after all there are about 7 billion people in the world, about 2 billion of them are children; where is this figure in the list of other causes of death among the world's young? Is the prevalence the same in different nations? And 42,000 is an extremely round number; it is surely not an exact count. What is the likely error? 10% or 50%? Or even 100%? A search of PLoS ONE leads to some answers and some troubling information.[1] The figure given is an annual figure, but it is only a speculative figure based upon a mathematical model that attempts to estimate the number of undiagnosed cases, among other things. And the article does not say that 42,000 die, it says "42,000 could die," a very different matter. The figure is not based on data; in fact, the PLoS ONE article notes "a profound lack of globally representative epidemio-logical data." It discusses a range of values (plus or minus 15%) found by reworking the model, but makes no statement about the range due to possible model failure. The article de-scribes its own model as "relatively crude." PLoS ONE gives us context; Science does not, and is grossly misleading to boot.

Measurements are only useful for comparison. The context supplies a basis for the comparison—perhaps a base-line, a benchmark, or a set of measures for intercompar-ison. Sometimes the baseline is implicit, based on general knowledge, as when the day's temperature is reported and can be related to local knowledge and past experience. But often, for example, in the case of childhood deaths from celiac disease, there is no such general knowledge. And in any case, more is needed in science: real data, clearly sourced, and a scale of measurement to assess magnitudes of difference. Is the difference notable or inconsequential?

One of the earliest examples of routine physical measurement was the trial of the Pyx, already discussed in Chapter 2. In that trial, even from the beginning—around 1100 CE—the baseline for weight was given by the indenture, a contractual baseline. The baseline for the fineness of the metal was given by the trial plate, a sample kept in the Tower of London specifically for that purpose. The trial of the Pyx also had a scale for assessing differences: the "remedy," giving what we now call a tolerance level. It was derived from negotiation, but there is no indication that it was in any formal way derived from data, from a formal assessment of the variability of the minting process. And, as we noted, there was a flaw in the way it was applied.

Arbuthnot and Significance Tests

In modern Statistics we use a probability measure as at least part of the assessment of differences, often in the form of a statistical test with roots going back centuries. The structure of a test is as an apparently simple, straightforward question:

Do the data in hand support or contradict a theory or hypothesis? The notion of likelihood is key to answering this question, and it is thus inextricably involved with the construction of a statistical test. The question addressed by a test would be answered by a comparison of the probability of the data under different hypotheses. In the earliest examples, only one probability was computed and the comparison was implicit.

John Arbuthnot was best known as a provocative writer who created the prototypical British character John Bull in a satire entitled The Law Is a Bottomless Pit, published in 1712. Arbuthnot was a close friend of Jonathan Swift and Alexander Pope; Pope wrote a famous satire as a letter to his friend, "An Epistle to Dr. Arbuthnot," in which, while criticizing Joseph Addison, he introduced the phrase, "damn with faint praise." Arbuthnot was also trained in mathematics and in medicine (from 1705 to 1714 he served as the personal physician to Queen Anne), and as a mathematician he made two notable contributions. The first was a pamphlet on probability published in 1692; it was largely a translation of a Latin tract by Christiaan Huygens published in 1657, but it was one of the earliest English publications on the subject. The second was a short note read to the Royal Society of London in 1710 and subsequently published in their Transactions. It was entitled,

"An Argument for Divine Providence, Taken from the Constant Regularity Observ'd in the Births of Both Sexes."[2] This note is often cited today as an early example of a significance test.

Arbuthnot's argument that the observed balance in the numbers of males (M) and females (F) could not be due to chance, and thus must be the effect of Divine Providence, was in two parts. The first was to show mathematically that if sex were assigned as if from the tosses of a fair two-sided die, an exact balance (or even a very close balance) was exceedingly improbable. He computed the chance of an exact balance in sex among two people (that is, the probability of the two being MF or FM, namely, $1/4 + 1/4 = 1/2$), among six people ($20/64 = 0.3125$), and among 10 people ($63/256 < 1/4$), and he stated that with logarithms this could be carried to a very great number of people and would clearly yield a very small probability. All this was correct: the chance that in $2n$ tosses of a fair coin the numbers of heads and tails will balance exactly is approximately $c/\sqrt{n}$, where $c = \sqrt{(2/\pi)} = 0.8$, as the table shows.

With less accuracy, Arbuthnot claimed that the chance would remain small if you widened the definition of balance from exact balance to approximate balance; in that instance, the question of what *approximate* meant was crucial,

Number of tosses	Probability of exact balance
2	0.50
6	0.31
10	0.25
100	0.08
1,000	0.025
10,000	0.008

and the mathematics needed to do the calculation was a few years in the future. But in any event it was the second argument he presented that has earned him a footnote in history.

Arbuthnot examined the excess of male births over female births in 82 years of the bills of mortality (see Figure 3.1), and he found that such a run would occur by chance only 1 time in 2^{82}, a chance too small to be admitted: 1/4,836,000,000,000,000,000,000,000.

Here the "random distribution," that is, sex being assigned in each case independently with equal probability, was set in comparison with the workings of Divine Providence, a hypothesis that would give a greater probability to there being more males than females in these data. Why? Because in view of the "external Accidents to which are Males subject (who must seek their Food with danger)," "provident Nature, by the disposal of its wise Creator, brings forth more Males than Females; and that in almost constant proportion."[3] Arbuthnot made no calculation for that alternative.

	Christened.			Christened.	
Anno.	Males.	Females.	Anno.	Males.	Females.
1629	5218	4683	1648	3363	3181
30	4858	4457	49	3079	2746
31	4422	4102	50	2890	2722
32	4994	4590	51	3231	2840
33	5158	4839	52	3220	2908
34	5035	4820	53	3196	2959
35	5106	4928	54	3441	3179
36	4917	4605	55	3655	3349
37	4703	4457	56	3668	3382
38	5359	4952	57	3396	3289
39	5366	4784	58	3157	3013
40	5518	5332	59	3209	2781
41	5470	5200	60	3724	3247
42	5460	4910	61	4748	4107
43	4793	4617	62	5216	4803
44	4107	3997	63	5411	4881
45	4047	3919	64	6041	5681
46	3768	3395	65	5114	4858
47	3796	3536	66	4678	4319

B b Chriſtened.

	Christened.			Christened.	
Anno.	Males.	Females.	Anno.	Males.	Females.
1667	5616	5322	1689	7604	7167
68	6073	5560	90	7909	7302
69	6506	5829	91	7662	7392
70	6278	5719	92	7602	7316
71	6449	6061	93	7676	7483
72	6443	6120	94	6985	6647
73	6073	5822	95	7263	6713
74	6113	5738	96	7632	7229
75	6058	5717	97	8062	7767
76	6552	5847	98	8426	7626
77	6423	6203	99	7911	7452
78	6568	6033	1700	7578	7061
79	6247	6041	1701	8102	7514
80	6548	6299	1702	8031	7656
81	6822	6533	1703	7765	7683
82	6909	6744	1704	6113	5738
83	7577	7158	1705	8366	7779
84	7575	7127	1706	7952	7417
85	7484	7246	1707	8379	7687
86	7575	7119	1708	8239	7523
87	7737	7214	1709	7840	7380
88	7487	7101	1710	7640	7288

3.1 (above and below) Arbuthnot's data. (Arbuthnot 1710)

Similarly, in a prize essay published in 1735, Daniel Bernoulli considered the surprising closeness of the planetary orbit planes of the five other known planets to that of the earth.[4] The six orbital planes were not in perfect agreement, but they did fall within a small angular difference: the inclines of their planes of orbit were all within 6° 54' of one another. Bernoulli judged this close of an agreement to be too unlikely to be admitted under a hypothesis of random distribution. In one of his calculations, he took 6° 54' as about 1/13 of 90°, and he judged the chance that the other five would incline within a given 6° 54' interval including the earth's orbit to be $(1/13)^5 = 1/371,293$. For Bernoulli, this gave the probability that all orbital planes would agree within the minimum angle needed to encompass them all.

Both Arbuthnot and Bernoulli had put a probability yardstick to a set of data, essentially using a principle Ronald A. Fisher would later articulate as a logical disjunction: "The force with which such a conclusion is supported is logically that of a simple disjunction: Either an exceptionally rare chance has occurred, or the theory of random distribution is not true."[5] If the data were not the result of a random distribution, some other rule must govern. In both Arbuthnot's and Bernoulli's cases, the likelihood comparison was implicit—it being taken for granted that at least one other hypothesis (Divine Providence or Newtonian dynamics) would yield a much higher probability for the observed data than the probability found under a hypothesis of "chance."

When the comparison problem is simple, when there are only two distinct possibilities, the solution can be simple as well: calculate one probability, and, if it is very small, conclude the other. At first blush, the problems of Arbuthnot and Bernoulli seemed of that type, but even then, difficulties intruded. With Arbuthnot's first discussion, this occurred when he granted that the balance need not be exact, but only approximate, raising the question, How close would be good enough? And so he went to the London birth data, where he found 82 years of more male than female births and could calculate that one probability, concluding "other." Arbuthnot's calculation had some aspects of a modern test, but it only treated an extreme situation: male births had exceeded female births in every one of the available 82 years. What might he have done if the excess appeared in only 81 of the 82 years? Would he have evaluated the probability of *exactly* 81 of 82 under a chance hypothesis? Or would he (as modern tests would tend to do) have found the probability of *at least* 81 of 82? Either of those probabilities is miniscule, but what about more intermediate cases such as 60 of the 82, or 48 of the 82, where the different approaches can give very different answers? We have no idea what he might have done.

This problem would be exacerbated if the data were recorded on a continuous scale (or approximately so), so that under most reasonable hypotheses *every* data value has a miniscule probability. In a population where an individual's sex at birth is equally likely to be male or female, and each sex is independently determined, then if there are 1,000,000 births, no possible number of male births has probability greater than 1/1,000. Does that mean that we reject the hypothesis of a natural random balance even if the data show a perfectly equal number of individuals for each of the sexes? Clearly the calculation of a single probability is not the answer to all questions. A probability itself is a measure and needs a basis for comparison. And clearly some restriction on allowable hypotheses is needed, or else a self-fulfilling hypothesis such as "the data are preordained" would give probability one to any data set.

Hume, Price, and Bayesian Induction

Not all likelihood arguments were explicitly numerical. A famous example is David Hume's argument against some of the basic tenets of Christian theology. In 1748, Hume published an essay, "Of Miracles," that he had composed somewhat earlier but withheld because of the stir he expected it to create.[6] Hume argued that no credence should be given to reported miracles, with that of the resurrection of Christ being a prime example. Hume characterized a miracle as "a violation of the laws of nature," and, as such, it was extremely improbable.[7] So improbable, in fact, that it was overwhelmed in comparison to the surely larger probability that the report of the miracle was inaccurate, that the reporter either lied or was simply mistaken.

Hume was correct in expecting a controversy, but he may not have anticipated the mathematical character of one response. It was just at that time, and probably in response to Hume, that Thomas Bayes wrote at least a good part of his famous essay. There is, in any event, no doubt that when Richard Price saw Bayes's essay through to publication in early 1764, Price considered his goal to be answering Hume's essay.[8] Price's intended title for Bayes's essay (and quite possibly Bayes's intention as well) has only recently come to notice: "A Method of Calculating the Exact Probability of All Conclusions Founded on Induction" (see Figure 3.2). This title is so bold that no text could possibly fully justify it. The article did present a mathematical treatment of this question: if an event has unknown probability p of occurring in each of n independent trials, and is found to have occurred in x of them, find the a posteriori probability distribution of p under the a priori hypothesis that all values of p are equally likely. This was the first appearance of a special case of Bayes's Theorem, and, as Price's next substantial publication showed, Hume was the intended target.

A   M E T H O D

OF CALCULATING

THE EXACT PROBABILITY

O F

All Conclufions founded on INDUCTION.

By the late Rev. Mr. THOMAS BAYES, F. R. S.

Communicated to the Royal Society in a Letter to

JOHN   CANTON, M. A. F. R. S.

A N D

Publifhed in Vol. LIII. of the Philofophical Tranfaƈtions.

With an APPENDIX by R. PRICE.

Read at the ROYAL SOCIETY Dec. 23, 1763.

L O N D O N:

Printed in the YEAR M.DCC. LXIV.

3.2 The title page chosen by Richard Price for the offprint of Bayes's essay. *(Watson 2013)*

In 1767, Richard Price's book *Four Dissertations* appeared, and a part of it directly rebutted Hume with explicit citation of Bayes's essay under the more provocative title.[9] It included Price's application of Bayes's Theorem (the earliest after the essay itself), an explicit calculation to show that a violation of what was seen as a natural law was not so un-likely as Hume argued. Hume had insisted that since the testimony for the existence of miracles was based solely upon experience, the defense of miracles needed to be so based as well. Price's argument went as follows. Suppose that the testimony in support of the natural law was that the same event (for example, a rising tide or a daily sunrise) had happened 1,000,000 times in a row with no exception. Consider this as observing $n = 1,000,000$ binomial trials, and finding that the number of miraculous exceptions was $X = 0$. Does this imply that $p =$ the probability of a miracle in the next trial is zero? No. Price used Bayes's Theorem to compute that, in this situation, the conditional probability of the chance of a miracle being greater than $1/1,600,000$ was $\mathrm{Prob}\{p > 1/1,600,000 | X = 0\} = 0.5353$, a better than 50% chance. Admittedly, $1/1,600,000$ was quite small, but hardly to the point of impossibility. To the contrary, with that as the chance of a miracle in a single trial, Price found that the chance that at least one miracle would occur in the next 1,000,000 trials, namely,

$$1.0 - (1,599,999/1,600,000)^{1,000,000} = 0.465,$$

is nearly half! The possibility of a miracle was much larger than Hume had supposed.

Bayes's article went essentially unnoticed for about half a century after publication, no doubt partly due to the uninteresting title it was given in the printed journal ("An Essay toward Solving a Problem in the Doctrine of Chances"), and it would not be until well into the twentieth century that posterior probabilities would play an important role in calibrating inferences, moving toward a realization of the bolder title. We will return to this topic in Chapter 5.

Laplacian Testing

Throughout the nineteenth century there continued to be a large number of ad hoc calculations of what could be called significance probabilities, generally in the spirit of Daniel Bernoulli's computation, using the data as defining the extreme limit of a set of values whose probability would then be found under some hypothesis of randomness.

Pierre Simon Laplace, in 1827, looked at a long series of barometric readings at the Paris Observatory for evidence of a lunar tide in the atmosphere and found an effect of $x = 0.031758$. He then calculated the chance, absent any real effect, of an effect no larger than that (in absolute value) to be 0.3617. This would correspond to a modern two-sided P-value of $1 - 0.3617 = 0.6383$, and he judged 0.3617 too small (that is, P was too large) to bear up as evidence for a tide. Laplace wrote,

> If this probability [0.3617] had closely approached unity, it would indicate with great likelihood that the value of x was not due solely to irregularities of chance, and that it is in part the effect of a constant cause which can only be the action of the moon on the atmosphere. But the considerable difference between this probability and the certainty represented as unity shows that, despite the very large number of observations employed, this action is only indicated with a weak likelihood; so that one can regard its perceptible existence at Paris as uncertain.[10]

Laplace's interpretation has stood the test of time; the effect of the lunar tide at Paris is too weak to detect with the observations then available. In contrast, he was able to find evidence of a seasonal effect on barometric variability (the mean change in pressure between 9 AM and 3 PM). Here he gave the modern P-value, stating that, in the absence of any seasonal effect, the chance of a deviation of the size he found or larger was calculated to be 0.0000015815, too small to be due to chance.

In 1840, Jules Gavarret compared the sex ratio at birth for "legitimate" births with the ratio for "illegitimate" births;[11] the fractions of male births were, respectively, 0.51697 and 0.50980, a difference of 0.00717. The numbers of births were large (1,817,572 legitimate and 140,566 illegitimate; see Figure 3.3), and he followed a guideline he attributed to Poisson and compared the difference with 0.00391, found as what we would now describe as $2\sqrt{2}$ ($=2.828$) times the estimated standard deviation of the difference, corresponding to an absolute deviation that would occur by chance with probability 0.0046. Since the observed difference was nearly twice that threshold, Gavarret interpreted this as showing that the deviation was larger than could be attributed to experimental variation. We of course would wish to remind him that the test could not tell us if the difference was due to social or biological factors.

In 1860, the American astronomer Simon Newcomb reconsidered an old problem from a new point of view.[12] Was it remarkable that six bright stars of the fifth magnitude should be found in a single, small square (1° square) of the celestial sphere, as is the case with the constellation Pleiades? Or might we expect this to occur with a reasonable probability even if the stars were scattered randomly in the heavens? The brightness of visible stars was then determined on a scale in which the dimmest stars visible were classed of the sixth magnitude, and the brighter over a range from the fifth on up. To a good approximation, there were $N = 1,500$ known stars of the fifth or greater magnitude, and the celestial sphere comprises 41,253 square degrees. Then

1824-1825.

Enfants légitimes.

$m = 939641 = $ le nombre de garçons,
$n = 877931 = $ le nombre des filles.
$\mu = 1817572 = $ le nombre des naissances.

D'où résulte que la chance moyenne de naissance d'un garçon en France dans l'état de mariage, est représentée par le rapport

$$\frac{m}{\mu} = \frac{939641}{1817572} = 0,51697$$

En poussant l'approximation jusqu'à la cinquième décimale.

Enfants illégitimes.

$m' = 71661 = $ le nombre des garçons,
$n' = 68905 = $ le nombre des filles.
$\mu' = 140566 = $ le nombre des naissances.

D'où résulte que la chance moyenne de naissance d'un garçon en France hors l'état de mariage, est représentée par le rapport

$$\frac{m'}{\mu'} = \frac{71661}{140566} = 0,50980$$

En poussant l'approximation jusqu'à la cinquième décimale.

3.3 Birth data from Gavarret. *(Gavarret 1840, 274)*

p=probability that a single random star would be in a specific square degree would be 1/41,253. Newcomb's analysis was original in that it treated the distribution of stars as a Poisson process, with $\lambda = Np = 1,500/41,253 = 0.0363$ as the rate of the spatial process, the expected number of stars per square degree. Then the chance s stars would be found in a specific square degree was

$$e^{-\lambda}\lambda^s/s!,$$

which for $s=6$ gives 0.000000000003. Since this referred to only a single, specified square degree, while the Pleiades had been selected for attention as the most densely packed square degree, Newcomb knew this was not the appropriate probability. So instead he found the expected number of the 41,253 square degrees that would be home to six stars, namely 41,253 times this small probability, 0.00000013, still a vanishingly small number. Actually, he knew this, too, was not the correct number, that what was wanted was the chance allowing that the square degree be adjusted by small shifts to encompass the most stars. But he thought that answer, which he was unable to compute, would not be much larger. He did note that one would have to enlarge the target space from 1 square degree to 27.5 square degrees in order that the expected number of such areas holding six stars would be 1.

A Theory of Likelihood

In the examples I have given, a growing sophistication can be seen, but there was also the beginning of a more formal theoretical development over that period. In the mid-1700s, several people began to express the combination of observations and the analysis of errors as a problem in mathematics. And some of these people, including Thomas Simpson (1757; see Figure 3.4), Johann Heinrich Lambert (1760; see Figure 3.5), Joseph-Louis Lagrange (1769), Daniel

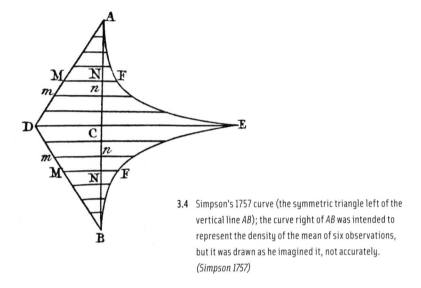

3.4 Simpson's 1757 curve (the symmetric triangle left of the vertical line *AB*); the curve right of *AB* was intended to represent the density of the mean of six observations, but it was drawn as he imagined it, not accurately. *(Simpson 1757)*

Bernoulli (1769, 1776; see Figure 3.6), Pierre Simon Laplace (1774 and later years; see Figure 3.7), and Carl Friedrich Gauss (1809), described a symmetric unimodal error curve or density as a part of the analysis, seeking to choose a summary of the data that was "most probable" with that curve in mind.

Some of these early analyses are recognizable as forerunners of what we now call maximum likelihood estimates.[13] These theories became increasingly refined: Laplace preferred a posterior median as minimizing the posterior expected error; Gauss also took a Bayesian approach in his first work

3.5 Lambert's 1760 curve. *(Lambert 1760)*

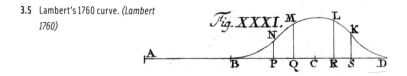

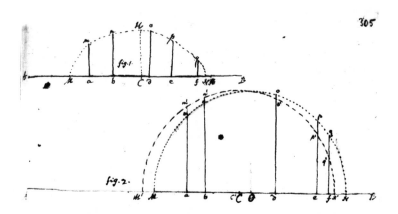

3.6 Daniel Bernoulli's 1769 curve (figure 1 in the diagram he drew) with a weighting function he used based upon that curve (figure 2 in his diagram). *(Bernoulli 1769)*

on this, also with a flat prior, and was led to the method of least squares as giving the "most probable" answer when errors were normally distributed (least squares had been published by Legendre four years before with no role for probability). But there was no full theory of likelihood before the twentieth century.

In the 1920s, building on some earlier work of Karl Pearson (including Pearson's influential introduction of the chi-

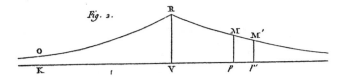

3.7   Laplace's 1774 curve, now called a double exponential density. *(Laplace 1774)*

square test in 1900), Fisher announced a remarkably bold and comprehensive theory.[14] If θ represents the scientific goal, and X the data, either or both of which could be multidimensional, Fisher defined the likelihood function $L(\theta|X)$ to be the probability or probability density function of the observed data X considered as a function of θ; we will follow custom and suppress the X in this notation as fixed by observation, writing $L(\theta) = L(\theta|X)$. He would take the θ that maximized $L(\theta)$, the value that in a sense made the observed data X the most probable among those values θ deemed possible, and he described this choice as the maximum likelihood estimate of θ. Thus far, except for the terminology, he was with Daniel Bernoulli, Lambert, and Gauss. But Fisher also claimed that when the maximum was found as a smooth maximum, by taking a derivative with respect to θ and setting it equal to 0, the accuracy (the standard deviation of the estimate) could be found to a good approximation from the curvature of L at its maximum (the second derivative), and the estimate so found expressed all the relevant information available in the data and could not possibly be improved upon by any other consistent method of estimation. As such, it would be the answer to all statisticians' prayers: a simple program for finding the theoretically best answer, and a full description of its accuracy came along almost for free.

Fisher's program turned out not to be so general in application, not so foolproof, and not so complete as he first thought. Rigorous proofs were difficult, and some counterexamples were found. Most of these were pathological—examples that were real but not troubling for practice. One exception to this was apparently discovered separately by Maurice Bartlett in 1937 and by Abraham Wald in 1938 (who communicated it to Jerzy Neyman in correspondence that same year).[15] Neyman, with Elizabeth Scott, published a simplified version of it a decade later, essentially as follows.[16] Suppose your data consists of n independent normally distributed pairs (X_i, Y_i), where in each pair X and Y are independent measures with the same expectation μ_i, but all Xs and Ys have the same variance σ^2. There are then $n+1$ quantities to be estimated. The maximum likelihood estimates of the μ_i are the pairs' means, $(X_i + Y_i)/2$, and the maximum likelihood estimate of the variance σ^2 is $\Sigma(X_i - Y_i)^2/4n$, an estimate whose expectation is $\sigma^2/2$, just half of what it should be. The difficulty arises because this is the average of the n separate estimates of variance for each sample of 2. In the normal case, the maximum likelihood estimate of variance for sample size m is biased, equaling the variance times $(m-1)/m$, which will be near 1 if m is large but is $\frac{1}{2}$ for $m=2$. Now think of this as a big data problem, in which the number of data recorded is about equal to the number of targets; the information in the overall sample must be spread over a huge number of targets and cannot do all parts of the job well. The example may be viewed as troubling for the application of maximum likelihood to big data, or encouraging: It does as well as could be expected for the means, and the problem with the variance is easily compensated for (just multiply by 2). But it does suggest caution in dealing with high-dimensional problems.

Despite these setbacks, Fisher's program not only set the research agenda for most of the remainder of that century; the likelihood methods he espoused or their close relatives have dominated practice in a vast number of areas where they are feasible.

While Fisher made much use of significance tests, framed as a test of a null hypothesis without explicit specification of alternatives, it fell to Neyman and Egon Pearson to develop a formal theory of hypothesis testing, based upon the explicit comparison of likelihoods and the explicit introduction of alternative hypotheses. The idea of testing, whether in the sense of Fisher or of Neyman and Pearson, has clearly been enormously influential, as testified to by its ubiquity and by the attention given to denouncing it in some quarters, or at least to denouncing some ways it has been implemented, such as an uncritical acceptance and routine use of a 5% level. The associated idea of likelihood as a way to calibrate our inferences, to put in a statistical context the variability in our data and the confidence we may place in observed differences, has become a pillar for much of the edifice of modern Statistics.

CHAPTER 4

· INTERCOMPARISON ·

Within-Sample Variation as a Standard

THE FOURTH PILLAR, INTERCOMPARISON, IS THE IDEA THAT statistical comparisons may be made strictly in terms of the interior variation in the data, without reference to or reliance upon exterior criteria. A vague sense of the idea is quite old, but the precise statement of what I have in mind only came in an 1875 article by Francis Galton. The scientific extensions of the idea in ways that make it radical and a mainstay of modern Statistics postdate Galton's article by 10, 33, and 50 years, in work by Francis Edgeworth, William Sealy Gosset, and Ronald A. Fisher, respectively.

In that 1875 article, "Statistics by Intercomparison," Galton presented a method of comparison with several desirable attributes, including that, in making comparisons, "we are able to dispense with standards of reference, in the common acceptance of the phrase, being able to create and afterwards indirectly to define them; . . . [they are] to be effected wholly by *intercomparison*, without the aid of any exterior standard."[1] That definition applies to the concept as later developed, but Galton's own use was limited to the use of percentiles, specifically (but not exclusively) the median and the two quartiles. These could be determined by simply ordering the data, without need for more complicated arithmetical calculations than counting, and served very well even in some cases where the measures were descriptive, ordered but not numerical. Indeed, Galton's initial use of percentiles came in 1869 in his book *Hereditary Genius*, where he used sets of biographical dictionaries to rank and compare talents in population groups, without any numerical measure of talent.[2] Not all of that book enjoys a high reputation today, but the statistical method was sound.

Gosset and Fisher's *t*

With historical hindsight, we can say that the first seed of a more mathematical use of intercomparison was sown in 1908 by an unlikely parent. Gosset had, since 1899, been employed as a chemist by the Guinness Company in Dublin. He had been trained at New College, Oxford, in mathematics (with a first in mathematical moderations in 1897) and in chemistry (with a first class degree in 1899), and he soon came to see that statistics could be useful to the brewery. In 1904–1905, he wrote a pair of internal memoranda (really in-house instruction texts) summarizing the uses of error theory and the correlation coefficient, based upon his reading of recent work from Karl Pearson's laboratory at University College London. One of Gosset's statements in the first of these memoranda expresses a wish to have a P-value to attach to data: "We have been met with the difficulty that none of our books mentions the odds, which are conveniently accepted as being sufficient to establish any conclusion, and it might be of assistance to us to consult some mathematical physicist on the matter."[3] That physicist was, of course, Pearson.

Guinness granted Gosset leave to visit Pearson's lab for two terms in 1906–1907 to learn more, and, while there, he wrote the article, "The Probable Error of a Mean," upon which his fame as a statistician rests (see Figure 4.1).[4]

The article was published in Pearson's journal Biometrika in 1908 under the pseudonym "Student," a reflection of Guinness's policy insisting that outside publications by employees not signal their corporate source. The article did not feature its potential applicability to quality control in brewing, and it would have been viewed at the time as an unremarkable

VOLUME VI   MARCH, 1908   No. 1

BIOMETRIKA.

THE PROBABLE ERROR OF A MEAN.

BY STUDENT.

Introduction.

ANY experiment may be regarded as forming an individual of a "population" of experiments which might be performed under the same conditions. A series of experiments is a sample drawn from this population.

Now any series of experiments is only of value in so far as it enables us to form a judgment as to the statistical constants of the population to which the experiments belong. In a great number of cases the question finally turns on the value of a mean, either directly, or as the mean difference between the two quantities.

If the number of experiments be very large, we may have precise information as to the value of the mean, but if our sample be small, we have two sources of uncertainty:—(1) owing to the "error of random sampling" the mean of our series of experiments deviates more or less widely from the mean of the population, and (2) the sample is not sufficiently large to determine what is the law of distribution of individuals. It is usual, however, to assume a normal distribution, because, in a very large number of cases, this gives an approximation so close that a small sample will give no real information as to the manner in which the population deviates from normality: since some law of distribution must be assumed it is better to work with a curve whose area and ordinates are tabled, and whose properties are well known. This assumption is accordingly made in the present paper, so that its conclusions are not strictly applicable to populations known not to be normally distributed; yet it appears probable that the deviation from normality must be very extreme to lead to serious error. We are concerned here solely with the first of these two sources of uncertainty.

The usual method of determining the probability that the mean of the population lies within a given distance of the mean of the sample, is to assume a normal distribution about the mean of the sample with a standard deviation equal to $s/\sqrt{n}$, where s is the standard deviation of the sample, and to use the tables of the probability integral.

Biometrika VI   1

4.1   The first page of "Student's" 1908 paper introducing what would become known as the *t*-test. *(Gosset 1908)*

product of the Pearson group—and in all respects but one it was just that. For over a century, scientists had routinely used the arithmetic mean in astronomy and described its accuracy in terms of the "probable error," or *p.e.*, defined as the median error for normally distributed data. In 1893, Pearson introduced the alternative scale of "standard deviation," or SD or σ, which was proportional to the p.e. (p.e. $\approx .6745\ \sigma$), and Pearson's usage soon became the norm. With large samples, statisticians would with no reluctance replace σ with $\sqrt{\frac{1}{n}\sum\left(X_i - \overline{X}\right)^2}$ (or, by Gauss's preference, $\sqrt{\frac{1}{n-1}\sum\left(X_i - \overline{X}\right)^2}$) when its value was not otherwise available. Gosset's goal in the article was to understand what allowance needed to be made for the inadequacy of this approximation when the sample was not large and these estimates of accuracy were themselves of limited accuracy. Specifically, he knew that $\overline{X}/\sigma$ had a normal distribution with mean zero and standard deviation $1/\sqrt{n}$ when the X_is were normal with mean zero. But what happened when σ was replaced by $\sqrt{\frac{1}{n}\sum\left(X_i - \overline{X}\right)^2}$? What was the distribution of $z = \overline{X}/\sqrt{\frac{1}{n}\sum\left(X_i - \overline{X}\right)^2}$? Fisher would later alter the scale to that of the now-familiar one-sample t-statistic, the relationship being $t = \sqrt{n-1}\,z$.

With a few good guesses that were unsupported by rigorous proof, and sound analysis based upon those guesses, Gosset derived what turned out to be the correct answer, what we now (on Fisher's scale) call Student's t distribution with $n-1$ degrees of freedom. There was some mathematical luck involved: Gosset implicitly assumed that the lack of correlation between the sample mean and the sample standard deviation implied they were independent, which was

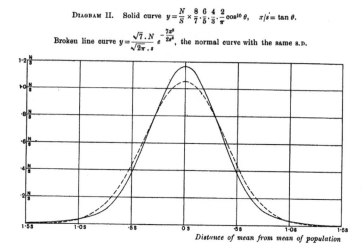

DIAGRAM II. Solid curve $y = \frac{N}{S} \times \frac{8}{7} \cdot \frac{6}{5} \cdot \frac{4}{3} \cdot \frac{2}{\pi} \cos^{10} \theta,$ $x/s = \tan \theta.$

Broken line curve $y = \frac{\sqrt{7} \cdot N}{\sqrt{2\pi} \cdot s} e^{-\frac{7x^2}{2s^2}},$ the normal curve with the same S.D.

4.2 Diagram from the 1908 paper comparing the normal and t (9 df) densities. *(Gosset 1908)*

true in his normal case but is not true in any other case. Figure 4.2 shows his distribution for z (solid line) with 9 degrees of freedom, compared to a normal density (dashed line) with the same standard deviation ($1/\sqrt{7} = .378$ on this scale). He noted the agreement was not bad, but for large deviations the normal would give "a false sense of security."[5]

The larger n, the closer his curve was to the normal. Gosset appended a table to permit significance probabilities to be calculated for n=4,5, . . . ,10, and he illustrated its use with a few examples, the most famous of which was the "Cushny-

SECTION IX. *Illustrations of Method.*

Illustration I. As an instance of the kind of use which may be made of the tables, I take the following figures from a table by A. R. Cushny and A. R. Peebles in the *Journal of Physiology* for 1904, showing the different effects of the optical isomers of hyoscyamine hydrobromide in producing sleep. The sleep of 10 patients was measured without hypnotic and after treatment (1) with D. hyoscyamine hydrobromide, (2) with L. hyoscyamine hydrobromide. The average number of hours' sleep gained by the use of the drug is tabulated below.

The conclusion arrived at was that in the usual dose 2 was, but 1 was not, of value as a soporific.

Additional hours' sleep gained by the use of hyoscyamine hydrobromide.

Patient	1 (Dextro-)	2 (Laevo-)	Difference (2-1)
1.	+ ·7	+ 1·9	+ 1·2
2.	− 1·6	+ ·8	+ 2·4
3.	− ·2	+ 1·1	+ 1·3
4.	− 1·2	+ ·1	+ 1·3
5.	− 1	− ·1	0
6.	+ 3·4	+ 4·4	+ 1·0
7.	+ 3·7	+ 5·5	+ 1·8
8.	+ ·8	+ 1·6	+ ·8
9.	0	+ 4·6	+ 4·6
10.	+ 2·0	+ 3·4	+ 1·4
	Mean + ·75	Mean + 2·33	Mean + 1·58
	S. D. 1·70	S. D. 1·90	S. D. 1·17

4.3 The Cushny-Peebles data from the 1908 paper. The "-1" in column 1 is a misprint for "-0.1." *(Gosset 1908)*

Peebles" data (see Figure 4.3). Working with the last column, the pairwise differences, he found $z = 1.58/1.17 = 1.35$; that is, the mean difference was 1.35 SDs from 0. This would give $t = \sqrt{n-1}\,(1.35) = 3(1.35) = 4.05$. This led him to say, "From the table the probability is .9985, or the odds are about 666 to 1 that 2 is the better soporific."[6] Student's t-test was born and had taken its first breath! We could find fault in the unwarranted Bayesian phrasing of the conclusion, the incorrect citation of the source (the article was published in 1905, not 1904), the misidentification of the drug (he mislabeled the columns, and, in fact, the data he copied were not for soporifics), and the inappropriate analysis (the individuals' data were in fact means of very different size samples and thus very different variances, and they were likely correlated by use of a common scaling). But at least the numerical work was clear and correct, and others would have no trouble in following the logic.

For present purposes, the important point is that the comparison, of the sample mean with the sample standard deviation, was made with no exterior reference—no reference to a "true" standard deviation, no reference to thresholds that were generally accepted in that area of scientific research. But more to the point, the ratio had a distribution that in no way involved σ and so any probability statements involving the ratio t, such as P-values, could also be made interior to the data. If the distribution of that ratio had varied with σ, the evidential use of t would necessarily also vary according to σ. Inference from Student's t was a purely internal-to-the-data analysis. This use of intercomparison had great power, by freeing itself from the need for such inputs. It also would open itself to criticisms of the type that were already common in 1919 and are undiminished today: that statistical significance need not reflect scientific significance.[7] Was the difference claimed of any practical significance as a means of inducing sleep? Gosset was silent on that. But while the potential for misleading statements remains an issue, the power that comes from the ability to focus on the data at hand is an undeniable benefit.

Gosset's article was all but ignored after publication. The journal was prominent, and a few surveys made routine reference to the article, but no one seems to have actually used the test in publication before the 1920s. In his *Tables for Statisticians and Biometricians*, published in 1914, Pearson included Gosset's test and table and gave the examples from the 1908 paper, including the uncorrected Cushny-Peebles data and the Bayesian conclusion.[8] But my search for even a single example of its use before 1925 has been fruitless. An afternoon spent in the Guinness Archives in Dublin, searching through the scientific memoranda from 1908 to 1924, revealed no examples: Gosset himself ignored the test in his practical work. There were several examples of the use of statistics, and mean differences were described in terms of how many standard deviations from zero they were, but there was no t-test and no reference to the paper in practice.

The paper had a profound influence nonetheless, all of it through the one reader who saw magic in the result. Fisher had probably read the paper around the time of his graduation from Cambridge in 1912. He had seen that there was no proof, but he had also seen that viewing the problem in terms of multidimensional geometry could yield an easy and fully rigorous proof. He sent a letter to Gosset (having somehow learned "Student's" true identity), explaining the proof. Gosset could not understand it; neither could Pearson when Gosset forwarded it to him. That letter was lost and probably never replied to. In 1915, Fisher included that proof in a short tour de force article in *Biometrika*, where he also found the distribution of a much more complicated statistic, the correlation coefficient r.[9]

Gosset's test still attracted no attention. By the early 1920s, when Fisher was working on agricultural problems at Rothamsted Experimental Station, he had seen that the mathematical magic that freed the distribution of Student's t from dependence on σ was but the tip of a mathematical iceberg. He invented the two-sample t-test and derived the distribution theory for regression coefficients and the entire set of procedures for the analysis of variance.

The historical impact of Gosset's work on statistical practice can be traced to its inclusion in Fisher's pathbreaking textbook, *Statistical Methods for Research Workers*, published in 1925.[10] Gosset's own paper presented a good idea, but only to the extent of a one-sample test, a species that is seldom useful other than as a test using pairwise differences as the sample. Fisher seized on the idea and expanded it to two and many samples, and it was there that the method's really powerful payoff became evident. Fisher's analysis of variance was really an analysis of variation, and he was able to pull variation apart in a way no one had even attempted before. Well, that is not quite true; Edgeworth had done something remarkable 40 years earlier.

Francis Edgeworth and Two-Way Analyses
for Variance Components

In the 1880s, Edgeworth undertook to extend the use of probability scales to the social sciences. As a part of this effort, he developed a method of analysis of statistical tables that anticipated a part of what Fisher would do later. At a meeting of the British Association for the Advancement of Science in Aberdeen in September 1885, Edgeworth presented his method in the context of two examples, one purposely fanciful, one more recognizably from social science.[11] For the first, he tabulated the number of dactyls (a metrical foot of a long syllable followed by two short syllables) in a segment of Virgil's *Aeneid* (see Figure 4.4). For the second, he took the death rates for eight years in six English counties from the registrar-general's report for 1883 (see Figure 4.5). In both, he found for each row and column the sum and mean, and what he termed the "fluctuation," or twice what

Æneid, XI, 1-75	Lines 1—5	6—10	11—15	16—20	21—25	26—30	31—35	36—40	41—45	46—50	51—55	56—60	61—65	66—70	17—75	Sums	Means	Fluctuations
First foot	3	3	5	5	4	4	2	2	2	1	2	4	3	2	4	46	3·06	2·8
Second ,,	1	4	0	3	3	3	5	2	2	4	3	1	2	3	2	38	2·5	3·2
Third ,,	1	2	4	2	5	2	1	2	2	2	0	2	2	0	1	28	1·86	3·1
Fourth ,,	2	2	1	0	3	1	2	0	2	1	1	2	1	1	0	19	1·26	1
Sums ..	7	11	10	10	15	10	10	6	8	8	6	9	8	6	7	131	8·68	10·0
Means ...	1·75	2·76	2·5	2·5	3·75	2·5	2·5	1·5	2	2	1·5	2·25	2	1·5	1·75	33	2·17	2·5 / 0·6
Fluctuations	1·5	2·5	9	7	3	3	5	2	0	3	3	2·5	1	3	2·5	208/48	3·0/3·2	

4.4   Edgeworth's analysis of data from Virgil's *Aeneid,* from 1885. Some numerical errors here and in Figure 4.5 are corrected in Stigler (1999), chapter 5. *(Edgeworth 1885)*

	1876.	1877.	1878.	1879.	1880.	1881.	1882.	1883.	Sums.	Means.	Fluctuations.
Berks............	175	172	187	186	181	153	169	166	1,389	173½	224
Herts	174	165	185	184	176	186	163	188	1,401	175	176
Bucks	182	171	186	195	179	162	177	183	1,435	179½	172
Oxford	179	182	194	183	180	169	167	166	1,420	177½	162
Bedford	196	174	203	195	198	171	181	184	1,502	188¼	246
Cambridge	173	177	190	191	187	165	171	181	1,435	179¼	158
Sums	1,079	1,041	1,145	1,134	1,101	986	1,028	1,068	8,582	1,073	1,138
Means	180	173½	191	189	183½	164	171	178	1,630	179	190/146
Fluctuations.	124	55	77	50	107	68	73	152	—	88/46	—

4.5   Edgeworth's analysis of data on county death rates, from 1885. *(Edgeworth 1885)*

we would now call the empirical variance, that is, $2\sum(X_i - \overline{X})^2/n$, for the appropriate row or column.

There was a subtext to Edgeworth's analysis. In both cases, the data were counts, either directly (Virgil) or scaled per ten thousand (death rates), and an approach being developed by Wilhelm Lexis at that time might have tried to base the analysis on an underlying version of binomial variation, an approach Edgeworth called "combinatorial." Edgeworth explicitly wanted to avoid such an assumption; his analysis was to be based solely on internal variation in the data—*intercomparison*, in Galton's terminology. Lexis's approach took binomial variation as an exterior standard. The temptation to use that simple coin-tossing model as a baseline was an old one—think of Arbuthnot and the data on sex at birth. But it came at a cost in more complicated situations. For the binomial with n trials and chance of success in a single trial p, the mean np and the variance $np(1 - p)$ are rigidly related, and not all data reflect that relation. Indeed, Arbuthnot's birth data is one of the rare cases that does, and most data since then are found to exhibit what analysts call over-dispersion: variation greater than the simple binomial, as would be the case if p varies randomly between trials. Edgeworth would avoid the strictures of Lexis. In modern terms, Edgeworth could handle the data whether they were binomial or not, as long as the variation was approximately normal.

Edgeworth framed his analysis as one of estimating what we would now call variance components. For example, grouping all death rates together, we could think of the overall "fluctuation" as a sum of three components: $C^2 + C_t^2 + C_p^2$, with the second component representing time (year-to-year) variation, the third representing place (county) variation, and the first representing random variation independent of time and place. An analyst wishing to compare rates over time for the same county would use the average or pooled fluctuation for rows (190 in Figure 4.5) as an estimate of $C^2 + C_t^2$ to assess accuracy, and, to compare counties in the same year, one would use the corresponding estimate (88 in the same figure) of $C^2 + C_p^2$. To estimate the random fluctuation C^2 he considered either looking at the difference, average row fluctuation minus fluctuation of the row of means, $190 - 146 = 44$, or the difference, average column fluctuation minus fluctuation of the column of means, namely, $88 - 46 = 42$. Because he was working numerically, not algebraically, he did not realize that, aside from errors in calculation, these should be exactly equal, both equal to $2SSE / IJ$, where I and J are the numbers of rows and column and SSE is the sum of squared residuals from fitting an additive model. Similarly, he could look for evidence that Virgil tended to use different metrical frequencies for different feet or for different segments of the poem.

Edgeworth's work was a series of missed opportunities, marred by numerical errors and algebraic clumsiness. His estimates were simple linear functions of the sums of squares in an analysis of variance, and some of his computations can be seen now as roughly equivalent to some of the F-statistics Fisher would later use for a similar analysis, but Edgeworth had no distribution theory to accompany them. When Fisher came to this problem in the mid-1920s, he clearly saw the full algebraic structure and the orthogonality that, with the mathematical magic of the multivariate normal distribution, permitted the statistical separation of row and column effects and enabled their measurement by independent tests of significance to be based only on the variation interior to the data.

With the increasing use of computers in the last half of the twentieth century came a greater use of computer-intensive procedures, including several that can be considered as using intercomparison. In the 1950s, Maurice Quenouille and then John W. Tukey developed a method of estimating standard errors of estimation by seeing how much the estimate varied by successively omitting each observation. Tukey named the procedure the jackknife. Related to this, several people have proposed variations under the name cross-validation, where a procedure is performed for subsets of the data and the results compared. In the late 1970s, Bradley Efron introduced a method he called the bootstrap that has become widely used, where a data set is resampled at random with replacement and a statistic of interest calculated each time, the variability of this "bootstrap sample" being used to judge the variability of the statistic without recourse to a statistical model.[12] All of these methods involve intercomparison in estimating variability.

Some Pitfalls of Intercomparison

Approaching an analysis with only the variation within the data as a guide has many pitfalls. Patterns seem to appear, followed by stories to explain the patterns. The larger the data set, the more stories; some can be useful or insightful, but many are neither, and even some of the best statisticians can be blind to the difference.

William Stanley Jevons was not the first to discern a business cycle in economic time series. He was also not the first to see a possible link between the business cycle and a similar cycle in sunspot activity. But his obsession with the idea in the late 1870s, even in the face of public and professional ridicule, exceeded that of his predecessors, and it was to stand as a caution to those who followed.

The close study of cycles has led to some of the greatest discoveries in the history of astronomy, but cycles in social science were of a different sort. Business cycles are indeed "periodic," but often with a shift—what one analyst of the results of horse races called "ever changing cycles." In the 1860s and 1870s, Jevons made careful studies of various series of economic data, and eventually he came to the conclusion that the regularity that a few others had seen was a genuine phenomenon: there was a regular business cycle, punctuated by a major commercial crisis about every 10.5 years.[13] He used data from a variety of published sources from the late 1700s to the 1870s, and eventually even pushed the record back to include the South Sea Bubble of 1720. When his first look at the data did not furnish an adequate crisis at close to the predicted time, he found he could make a case for at least a minor crisis at the appropriate time, usually after a thorough probing of the records. But what was the cause?

Much earlier, William Herschel had suggested that the regularity of the major bursts in sunspot activity could be related to the business cycle, but those who pursued that suggestion were hampered by a mismatch: the sunspot cycle peaked each 11.1 years—or so it was believed—and, after a few decades, that 11.1 and the 10.5 for business cycles got progressively out of synch. But when research by J. A. Broun in the mid-1870s revised the sunspot cycle length from 11.1 to 10.45, Jevons's interest in the matter turned into an obsession. He even "improved" upon the sunspot series: when there was a gap in the series, usually a minor maximum could be found that the eager Jevons could accept.

Some connection seemed undeniable, and when Jevons also found an approximation to cycles of about the same length in price statistics for grain in Delhi, the case was made. That solar activity could affect meteorology was plausible, and when the effect was pronounced, as seemed to be the case in India, there could be an effect on trade—after all, the British crisis of 1878 was precipitated by the failure of the City of Glasgow Bank, a failure itself attributed to a recent Indian famine that had depressed trade. This theory had the bonus of explaining the lag in crises, as the British crises seemed to follow the Delhi price series by a few years. Figure 4.6 is from Jevons's last paper, published in *Nature* in July 1882 (he died on August 13, 1882, in an accidental drowning at age 46).[14] It shows the sunspot series ("Wolf's numbers"), the Delhi price of grain ("corn"), and indicates the major British commercial crises as Jevons dated them. Without being aware of the sifting and seeking that went into choosing and recalculating these data, the impression is striking, even if not compelling to all viewers.

Jevons faced ridicule at the time. When he stated at a Statistical Society of London meeting in 1879 that he expected

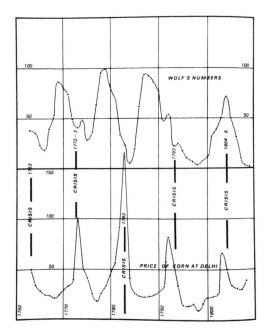

4.6 Jevons's chart showing the relationship of sunspot activity and commercial crises, from 1882. *(Jevons 1882)*

economic recovery in the next year or two, "provided, in-deed, that the sun exhibited the proper number of spots," the audience's amusement was not suppressed.[15] That same year the society's *Journal* ran two short anonymous pieces, one on how high counts of sunspots favored Cambridge over Oxford in the annual boat race, and another on a supposed link between death rates and the motions of Jupiter. In 1863, Galton had written, "Exercising the right of occasional suppression and slight modification, it is truly absurd to see how plastic a limited number of observations become, in the hands of men with preconceived ideas."[16] Of course, even totally benign time series can exhibit deceptive patterns. In 1926, in a paper provocatively titled, "Why Do We Sometimes Get Nonsense-Correlations between Time-Series?," G. Udny Yule showed how simple autoregressive series tend to show periodicity over limited spans of time.[17] He did not mention Jevons, possibly out of kindness.

CHAPTER 5

· REGRESSION ·

Multivariate Analysis, Bayesian Inference, and Causal Inference

CHARLES DARWIN HAD LITTLE USE FOR HIGHER MATHEMATICS. He summed up his view in 1855 in a letter to his old friend (and second cousin) William Darwin Fox with a statement that Karl Pearson later made famous: "I have no faith in anything short of actual measurement and the Rule of Three."[1] In 1901, Pearson adopted this as the motto for the new journal *Biometrika*, and, in 1925, when he founded the *Annals of Eugenics*, he went even further: he placed it on the title page of each and every issue (see Figure 5.1). It was as close to an endorsement of mathematics as Pearson could find in Darwin's writings.

Darwin was surely right in valuing actual measurement, but his faith in the Rule of Three was misplaced. The Rule

ANNALS OF EUGENICS

A JOURNAL FOR THE SCIENTIFIC STUDY
OF
RACIAL PROBLEMS

EDITED BY
KARL PEARSON
ASSISTED BY
ETHEL M. ELDERTON

VOL. I (1925–1926)

I have no Faith in anything short of actual Measurement and the Rule of Three.
CHARLES DARWIN

ISSUED BY THE
FRANCIS GALTON LABORATORY FOR NATIONAL EUGENICS
UNIVERSITY OF LONDON
AND PRINTED AT THE
UNIVERSITY PRESS, CAMBRIDGE

5.1 The title page of the first issue
of the *Annals of Eugenics*.

of Three that Darwin cited would have been familiar to every English schoolboy who had studied Euclid's book 5. It is simply the mathematical proposition that if $a/b = c/d$, then any three of a, b, c, and d suffice to determine the fourth. For Darwin, this would have served as a handy tool for extrapolation, just as it had for many others before him (see Figure 5.2). In the 1600s, John Graunt and William Petty had used such ratios to estimate population and economic

28   GOLDEN RULE.

II. EXAMPLES OF WEIGHTS, MEASURES, &c.

TROY WEIGHT.   APOTHECARIES WEIGHT.
lb oz dwt   lb oz dr sc gr
4) 13   1 15 (6) 2   5   3   0 19 (
AVOIRDUPOIS WEIGHT.   AVOIRDUPOIS WEIGHT.
c   qr lb   lb oz dr
7) 75   1 12 (9) 5   3 14 (
LONG MEASURE.   LONG MEASURE.
miles fur pls   yds feet inc
11) 58   5 12 (12) 150 1 .7 (
CLOTH MEASURE.   LAND MEASURE.
yds qrs nls   ac ro pls
17) 31   2   3 (26) 17 3 17 (

See more of Compound Division under Rule II of *Rules of Practice.*

GOLDEN RULE, OR RULE-OF-THREE.

THE RULE-OF-THREE is that by which a number is found, having to a given number the same proportion which is between two other given numbers. For this reason it is sometimes named the *Rule of Proportion.*

It is called the Rule-of-Three, because in each of its questions there are given three numbers at least. And because of its excellent and extensive use, it is often named the *Golden Rule.*

For the stating, or rightly placing down the three given numbers, observe the following

RULE.

1. Write down the number which is of the same kind with the answer or number required.

2. Consider whether the answer ought to be greater or less than this number; then write respectively the greater or less of the two remaining numbers on the right of it for the third, and the other on the left for the first number or term.

3. Multiply the second and third terms together, divide the product by the first, and the quotient will be the answer.

Note 1. When you can conveniently multiply and divide as in Compound Multiplication and Division, it is best so to do.

2. But if not, reduce the compound terms to the lowest name mentioned in them, and the first and third to the same name, if they be not so already; then will the answer be of the same name with the 3d term.

3. When there happens to be a remainder after division, reduce it to the name next below the last quotient, and divide by the same divisor.

GOLDEN RULE.   29

so shall the quotient be so many of the said next name; do t is as long as there is any remainder, or till you have reduced it to the least name, and all the quotients together will be the answer.

4. If the 1st term, and either the 2d or 3d, can be divided by any number, without remainder, let them be divided, and the quotients used instead of them.

5. There are four other methods of operation besides the general one above delivered, any of which, when possible, performs the work much shorter than it. They are thus.—

First, Divide the 2d term by the first, multiply the quotient by the 3d, and the product will be the answer.

Second, Divide the 3d term by the first, multiply the quotient by the 2d, and the product will be the answer.

Third, Divide the 1st term by the 2d, divide the 3d by the quotient, and the last quotient will be the answer.

Fourth, Divide the 1st term by the 3d, divide the 2d by the quotient, and the last quotient will be the answer.

For an example, let it be proposed to find the value of 14 oz 8 dwt of gold, at 3l 19s 11d an ounce.

EXPLANA. Having stated the three terms by the general rule, as here annexed, the 3d term is reduced to pence, and the 3d to dwts, these being their lowest denominations, as directed in the 2d note. The 1st term is also reduced to dwts, that it may agree with the third, by the same note. The 2d term is then multiplied by the 3d, and the product divided by the 1st, according to the general rule, when the answer comes out 13809 pence, and 12 remaining over; which remainder being reduced to farthings, and these divided by the same divisor 20, by the third note, the quotient is 2 farthings, and 8 remaining.—Lastly, the pence are divided by 12 to reduce them to shillings, and these again by 20 for pounds, when the final sum comes out 57l 10s 9d 2q, for the answer.

But this question will also serve to illustrate some more of the notes, by means of which it can be easier solved than by the general rule as above given; for having stated it as before, and as here again annexed, and reduced the 1st and 3d terms to dwts, divide them each by 4, and use the quotients 5 and 72 instead of them, as by the 4th note.

                                          oz   l s d   oz dwt
                                       1 : 3 19 11 :: 14 8
                                          20 20   20
                                          ——   ——   ———
                                          20   79   288
                                               12
                                               ——
                                               959
                                               288
                                               ——
                                               7672
                                               7672
                                               1918
                                               ————
                                          2,0)27619,2
                                          13809 12/20 pence, or
                                          12)13809d 2 8/20 qr
                                          2,0)115,0s 9d 2 8/20 qr
                                          Ans. 57l 10s 9d 2 8/20 q.

                                          oz   l s d   oz dwt
                                       1 : 3 19 11 :: 14 8
                                          20   20
                                          4)20 31 19 4   4)288
                                            5   9   72

                                          5)28: 14 0
                                          Ans. 57l 10s 9d 2 8/20 q.

B 3

activity; in the 1700s and early 1800s, so, too, did Pierre Simon Laplace and Adolphe Quetelet.

Neither Darwin nor anyone before him realized what weak analytical support the Rule of Three provides. The rule works well in prorating commercial transactions and for the mathematical problems of Euclid; it fails to work in any interesting scientific question where variation and measurement error are present.[2] In such cases, the Rule of Three will give the wrong answer: the results will be systematically biased, the errors may be quite large, and other methods can mitigate the error. The discovery of this fact, three years after Darwin's death, is the fifth pillar. The discoverer was Darwin's cousin, Francis Galton, who announced the surprising discovery of this far-reaching phenomenon in an address in Aberdeen, Scotland, on September 10, 1885, where he christened it "regression."[3] The basic conception would enable major lines of development in Statistics over the half century following 1885. The story of the discovery is revealing, but, before moving to that account, it may be useful to explain what Euclid's error was and how it could have gone unnoticed for thousands of years.

Let us take but one of the examples Galton considered, a problem in anthropology of a sort that remains common today.[4] A partial skeleton of a man is found, little more than a complete thigh bone of length T, and the anthropologist would like to know the man's height H. The anthropologist has several comparable complete skeletons, giving him a set of pairs (T_i, H_i) from which he has calculated the arithmetic means, m_T and m_H. His plan is to use the Rule of Three, inferring the unknown H from these means, the known T and the relationship $m_T/m_H = T/H$. If the relationship were as mathematically rigorous as those considered by Euclid, with a constant ratio for all T_i/H_i, this would work. But here, as in all scientifically interesting problems, there is variation, and the regression phenomenon Galton discovered renders the Rule of Three inappropriate. At one extreme, if T and H vary but are uncorrelated, the best estimate of H ignores T; it is m_H. Only if they are perfectly correlated is Euclid's answer the correct one. In intermediate cases, Galton found there were intermediate solutions, and, curiously, that the relation for predicting H from T differs markedly from that for predicting T from H, and neither one agreed with Euclid.

The Road from Darwin to Galton's Discovery

Darwin's theory of the origin of species was incomplete as published in 1859, and it remained incomplete on the day he died in 1882. All theories are incomplete, in that more always remains to be done once they are advanced, and in that sense the more fertile a theory is, the more incomplete it is. But Darwin's theory was incomplete in a more fundamental way: there remained a problem with the argument that, had it been widely noted, could have caused difficulty. It was a subtle problem, and a full appreciation and articulation only came when a solution was found by Galton, three years after Darwin died.[5]

The problem involved the fundamental structure of Darwin's argument. In order to make the case for evolution by natural selection, it was essential to establish that there was sufficient within-species heritable variability: a parent's offspring must differ in ways that are inheritable, or else there can never be a change between successive generations. The first chapters of the *Origin of Species* established that in a most convincing manner, for both domestic and natural populations of plants and animals.[6] In so doing, Darwin inadvertently created a problem, an apparent contradiction within the theory.

Evidently, only two readers noticed the problem during Darwin's lifetime: the engineer Fleeming Jenkin, in a book review in 1867, and Galton a decade later. Jenkin's review recognized only one part of the problem and distracted from that part by presenting a set of other, unrelated issues.[7] In 1877, Galton fully articulated the matter and posed it as the

Generation 0

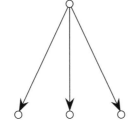

Generation 1

5.3 Variation produced in a single generation's offspring.

serious challenge it was.[8] Galton's formulation can be summarized graphically. Darwin had convincingly established that intergenerational transfers passed heritable variation to offspring (see Figure 5.3).

The same parent would produce offspring with different, heritable traits. Although Galton considered the question in general, it may help to think of the trait as stature—adult height—a trait Galton studied extensively, with female heights scaled up by 1.08 to allow for the known sex difference in stature. But if there was increased variation from parent to child, what about succeeding generations? Would not same pattern continue, with variation increasing in each successive generation? (See Figure 5.4.)

But increasing variation is not what we observe in the short term, over a sequence of generations. Within a species, the population diversity is much the same in successive generations (see Figure 5.5). Population dispersion is stable over the

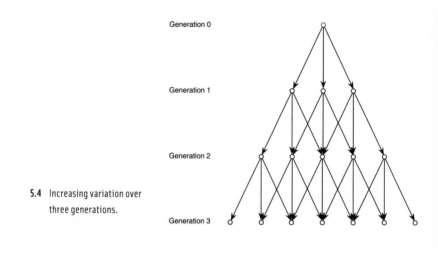

5.4 Increasing variation over three generations.

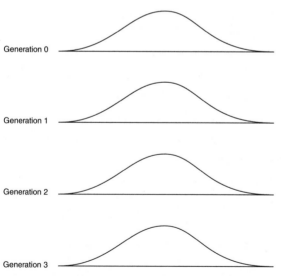

5.5 Population diversity is stable over three generations.

short term; indeed this stability is essential for the very definition of a species.

This year's crops show about the same diversity of sizes and colors as last year and the year before, both in nature and (absent aggressive breeding control) in domestic production. And any human population shows the same variation in size from generation to generation, absent marked changes in diet.

What Galton had in view was not the long-term evolution of species, where he was convinced that significant change had and would occur for reasons Darwin had given. His worry was short term; he was concerned about the implications of Darwin's theory even for what Galton called "typical heredity," when the timescale was such that one might expect at least an approximate equilibrium, absent any sudden change in environment. In even approximate equilibrium, the variability Darwin both required and demonstrated existed was in conflict with the observed short-term stability in populations. Darwin's model wouldn't work unless some force could be discovered that counteracted the increased variability yet also conformed with heritable intergenerational variation. Galton worked for a decade before he discovered that force, and, in effect, his success saved Darwin's theory.

Galton's solution was remarkable in being purely mathematical, albeit framed in terms of a series of analogue models. As such, it may be unique in early biological science. William Harvey's discovery of the circulation of blood was based upon arithmetical calculations, but it was predominantly empirical. A number of earlier scientists (for example, Lorenzo Bellini and Archibald Pitcairne) tried to create a mathematical medicine, but with no noted successes.[9] Galton was, in essence, able to establish some of the practical consequences of Mendelian genetics without the benefit of knowing any genetics, nearly two decades before Mendel's work was rediscovered.

Galton started in 1873 with the quincunx, a machine he devised to express intergenerational variability in which lead shot fell through a rake of rows of offset pins. The shot were deflected randomly left or right at each row, coming to rest at the bottom in one of several compartments (see Figure 5.6).

In 1877, he enlarged this idea to show the effect of this variability upon successive population distributions. In Figure 5.7, the top level represents the population distribution—say, of the stature in the first generation, with smaller stature on the left and larger on the right, in a roughly bell-shaped normal distribution.

In order to maintain constant population dispersion, he found it necessary to introduce what he called "inclined

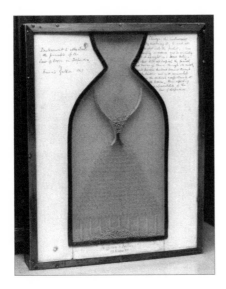

5.6 The original quincunx,
constructed in 1873 for use in
an 1874 public lecture. The lead
shot at the bottom give the
impression of a bell-shaped
curve. *(Stigler 1986a)*

chutes" to compress the distribution before subjecting it to
generational variability. In the middle, he showed schemat-
ically the effect of that variability upon two representative
groups of like stature, one in the middle and one to the right.
Each of these would produce a small distribution of offspring immediately below, shown as a little normal curve
proportional in area to the size of the group that produced
it. He computed exactly what the inclination of the chutes
must be (he called it a coefficient of reversion) in order for
intergenerational balance to be preserved, but he was
pretty much at a loss to explain why they were there. The

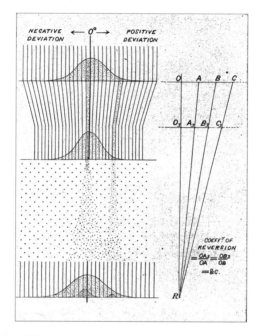

5.7 Galton's 1877 version of the quincunx, showing how the inclined chutes near the top compensate for the increase in dispersion below to maintain constant population dispersion, and how the offspring of two of the upper-level stature groups can be traced through the process to the lowest level. *(Galton 1877)*

best he could do in 1877 was to suggest that perhaps they represented a lesser propensity to survive—decreased fitness—for those furthest from the population mean. It was a faute de mieux excuse that to give an exact balance would seemingly have required a level of coincidence that even

Hollywood would not accept in a plot, and he did not mention it again.

To understand the solution Galton finally arrived at, and the wonderful device that carried him there, consider Figure 5.8 (focusing first on the left panel), an embellishment of one he published in 1889.[10] It shows a quincunx that has been interrupted in the middle (A). The lead shot are now stopped halfway down; the outline they would have produced had they not been interrupted is shown at the bottom (B). The two outlines of distribution at levels A and B are similar; they differ only in that the midlevel (A) is more crudely drawn (it is my addition) and is more compact than the lower level (B), as would be expected, since the shot at level A are only subjected to about half the variation.

Galton observed the following paradox. If you were to release the shot in a single midlevel compartment, say, as indicated by the arrow on the left panel, they would fall randomly left or right, but on average they would fall directly below. Some would vary to the left, some to the right, but there would be no marked propensity one way or the other. But if we look at, say, a lower compartment on the left, after all midlevel shot have been released and permitted to complete their journey to level B, and ask where the residents of that lower compartment were likely to have descended from, the answer is not "directly above." Rather, on average,

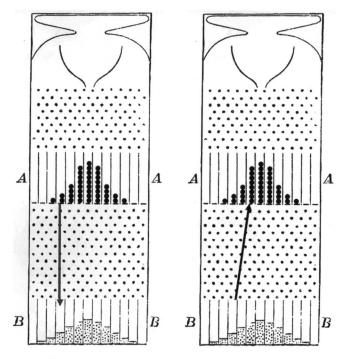

5.8 An embellished version of an 1889 diagram. The left panel shows the average
final position of shot released from an upper compartment; the right panel shows
the average initial position of shot that arrived in a lower compartment. *(Galton 1889)*

they came from closer to the middle! (see Figure 5.8, right
panel). The reason is simple: there are more level A shot in the
middle that could venture left toward that compartment than
there are level A shot to the left of it that could venture right
toward it. Thus the two questions, asked from two different

standpoints, have radically different answers. The simple reciprocity we might naively expect is not to be found.

Consider the relationship of Galton's quincunx to the data he gathered. Figure 5.9 shows Galton's table giving a cross-classification of the adult heights of 928 children from 205 sets of parents. The parents' heights are summarized as "mid-parent" heights, the average of the father's and mother's heights after the latter is scaled up by 1.08.[11] The female children's heights are similarly scaled up by 1.08. Look at the column "Total Number of Adult Children." Think of this as the counts of group sizes at the A level of the quincunx, corresponding to the groups described by the leftmost column of labels. The rows of the table give the history of the variability of the offspring within each group. For example, there were six mid-parents in the row of height labeled 72.5 inches; they produced 19 children whose adult heights ranged from 68.2 inches to "above," in a pattern akin to one of the little normal curves Galton showed at the bottom of Figure 5.7. Each row thus shows (in principle) such a little normal curve, and the row "Totals" then gives the summed counts; that is, the counts for the compartments shown at the bottom level of the quincunx (level B in Figure 5.8).

If people's heights behaved just like a quincunx, then the offspring should fall straight down from the mid-parents. The right-hand column of the table ("Medians") gives the

NUMBER OF ADULT CHILDREN OF VARIOUS STATURES BORN OF 205 MID-PARENTS OF VARIOUS STATURES.

(All Female heights have been multiplied by 1·08).

Heights of the Mid-parents in inches	Below	62·2	63·2	64·2	65·2	66·2	67·2	68·2	69·2	70·2	71·2	72·2	73·2	Above	Adult Children	Mid-parents	Medians
Above	··	··	··	··	··	··	··	··	··	··	··	··	3	··	4	5	··
72·5	··	··	··	··	··	··	··	1	2	1	··	1	3	4	19	6	72·2
71·5	··	··	··	··	1	3	4	3	5	10	4	9	2	2	43	11	69·5
70·5	1	··	··	··	··	1	3	12	18	14	7	4	3	3	68	22	69·5
69·5	··	··	1	16	4	17	27	20	33	25	20	11	4	5	183	41	68·9
68·5	1	··	7	11	16	25	31	34	48	21	18	4	3	··	219	49	68·2
67·5	··	3	5	14	15	36	38	28	38	19	11	4	··	··	211	33	67·6
66·5	··	3	3	5	2	17	17	14	13	4	··	··	··	··	78	20	67·2
65·5	1	··	9	5	7	11	11	7	7	5	2	1	··	··	66	12	66·7
64·5	1	1	4	4	1	5	5	··	2	··	··	··	··	··	23	5	65·8
Below	1	··	2	4	1	2	2	1	1	··	··	··	··	··	14	1	··
Totals	5	7	32	59	48	117	138	120	167	99	64	41	17	14	928	205	··
Medians	··	66·3	67·8	67·9	67·7	67·9	67·7	68·3	68·5	69·0	69·0	70·0	··	··	··	··	··

NOTE.—In calculating the Medians, the entries have been taken as referring to the middle of the squares in which they stand. The reason why the headings run 62·2, 63·2, &c., instead of 62·5, 63·5, &c., is that the observations are unequally distributed between 62 and 63, 63 and 64, &c., there being a strong bias in favour of integral inches. After careful consideration, I concluded that the headings, as adopted, best satisfied the conditions. This inequality was not apparent in the case of the Mid-parents.

5.9 Galton's data on family heights. (Galton 1886)

median height for the children for each group of mid-parents (each little-normal-curve-like group, evidently computed from ungrouped data). Galton noticed that these medians do not fall immediately below; instead, they tend to come closer to the overall average than that—a definite sign that the inclined chutes must be there! Invisible, to be sure, but they were performing in some mysterious way the task that his 1877 diagram had set for them. He presented a diagram (Figure 5.10) that shows this clearly.

Looking at the column medians, he noticed the same phenomenon there: each height group of children had an average mid-parent closer to the middle ("mediocrity") than they were. Galton also expected this—there were more mid-parents of mediocre height who could produce exceptional offspring than there were more extreme mid-parents who could produce less extreme children. But how did the chutes work—what *was* the explanation?

By 1885, Galton had one more piece of evidence, evidence that shined a new light on the phenomenon.[12] From a study he had made the previous year, he had data on a number of families, and he had thought to look at the data on brothers in the same manner as he had studied mid-parents and their children. The results were extraordinarily similar: the pattern of associations was the same (tallness ran in families), but most striking was that here, too, he found "regression." Look

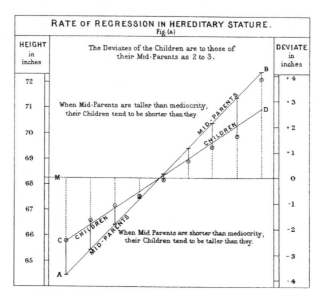

5.10   Here Galton plots numbers from the leftmost and rightmost columns of Figure 5.9, showing the tendency for children's heights to be closer to the population average than a weighted average of those of their mid-parents—a "regression towards mediocrity." *(Galton 1886)*

at the right-hand column of Figure 5.11: the medians here, too, are systematically closer to "mediocrity" than might have been expected.

This was extraordinary for the simple reason that, among the brothers in his table, there was no directionality—neither brother inherited his height from the other. There was no directional flow of the sort he had sought to capture with

Relative number of Brothers of various Heights to Men of various Heights, Families of Five Brothers and upwards being excluded.

Heights of the men in inches.	Heights of their brothers in inches.													Total cases.	Medians.
	Below 63	63·5	64·5	65·5	66·5	67·5	68·5	69·5	70·5	71·5	72·5	73·5	Above 74		
74 and above	1	1	..	..	..	..	..	1	1	..	5	3	12	24	—
73·5	..	..	..	..	..	1	3	4	8	3	3	2	3	27	71·1
72·5	..	..	..	..	1	1	6	5	9	9	8	3	5	47	70·2
71·5	..	1	..	1	2	8	11	18	14	20	9	4	..	88	69·6
70·5	..	..	1	1	7	19	30	45	36	14	9	8	..	171	69·5
69·5	..	1	2	1	11	20	36	55	44	17	5	4	1	198	68·7
68·5	..	1	5	9	18	38	46	36	30	11	6	3	2	203	67·7
67·5	2	4	8	26	35	38	38	20	18	8	1	1	..	199	67·0
66·5	4	3	10	33	28	35	20	12	7	2	1	..	..	155	66·5
65·5	3	3	15	18	33	26	8	2	1	1	..	..	..	110	65·6
64·5	3	8	12	15	10	8	5	2	1	..	..	..	..	64	
63·5	5	2	8	3	3	4	1	1	..	1	..	..	1	20	
Below 63	5	5	3	3	4	2	..	..	..	..	..	..	1	23	
Totals	23	29	64	110	152	200	204	201	169	86	47	28	25	1329	

5.11 Galton's 1886 data on brothers' heights. (Galton 1886)

his various quincunxes. The brothers' data were markedly symmetrical—aside from typos, he must have included each pair of brothers twice, once in each order. The "below 63 and 74 and above" pair given in the upper left corner and in the botton right corner of Figure 5.11 must be the same two individuals. How could "inclined chutes" play a role there? Indeed, even inheritance seems ruled out. It must have been clear to Galton that the explanation must be statistical, not biological.

Galton went back to the mid-parent and children data and smoothed the counts, averaging together the counts in groups of four cells and rounding, to make the pattern stand out better. He could see a rough elliptical contour emerging around the densest portion of the table. He wrote out a mathematical description of the actions of the quincunx (with mid-parent population as a normal distribution and each subgroup's offspring as a normal distribution with the same dispersion) and, with some help from a Cambridge mathematician, found a theoretical representation for the table— what we now recognize as the bivariate normal density, with the major and minor axes, and more importantly, the two "regression lines" (see Figure 5.12).[13] One was the theoretical line describing the expected height of the offspring as a function of the mid-parent's height (the line ON); the other was the expected height of the mid-parent as a function of the child's height (the line OM).

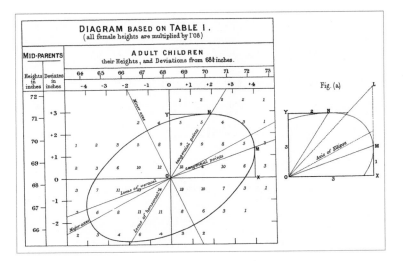

5.12 Galton's 1885 diagram showing the two regression lines *ON* and *OM*. The "Rule of Three" line is given in the small diagram on the right as the line *OL*. (*Galton 1886*)

The nature of the statistical phenomenon was becoming clear. Since the lines, whether the theoretical version in terms of the bivariate density or the numerical version in terms of the table, were found by averaging in two different directions, it would be impossible for them to agree unless all the data lay on the table's diagonal. Unless the two characteristics had correlation 1.0 (to use a term Galton introduced in late 1888 with these data in mind), the lines had to differ, and each had to be a compromise between the perfect correlation case (the major axis of the ellipse) and the zero correlation case (the horizontal [respectively, vertical] lines through the center). Interestingly, Galton's diagram also shows the line that the Rule of Three would give. It is the line OL in the right-hand panel, and it neither agrees with a regression line nor has any particular statistical interpretation. In this case, it is simply the 45° diagonal, reflecting the equality of the mean statures of the mid-parents and children populations.

Galton's Interpretation

In the book *Natural Inheritance* that Galton published in 1889 summarizing this research, he expressed his idea in words, suggesting that if P is the mean stature for the relevant population, then, given the stature of one brother, "the unknown brother has two different tendencies, the one to resemble the known man, and the other to resemble his race. The one tendency is to deviate from P as much as his brother, and the other tendency is not to deviate at all. The result is a compromise."[14] In modern terminology, we could represent the statures of the two brothers by $S1$ and $S2$, where each stature consists of two components, $S1 = G + D1$, $S2 = G + D2$, where G is an unobserved persistent component common to both brothers (a genetic component that they hold in common with each other) and D1 and D2 are unobserved transitory or random components, uncorrelated with G and with each other. Galton's P would represent the mean of all the G in the population.

We could then articulate the idea of regression as a selection effect. If we observe the stature S1 of the first brother as deviating above the population average P, on average, S1 likely deviates for some balance of two reasons: because an individual's G varies somewhat above P and because D1 varies somewhat above zero. When we turn to the second brother, his G will be the same as his brother's but, on average, the contribution of D2 will be nil, so S2 will be expected to be above P, but only to the extent of G, not G+D1, and thus not so much as S1. And the same argument works with S1 and S2 reversing positions.

The Solution to Darwin's Problem

Galton had discovered that regression toward the mean was not the result of biological change, but rather was a simple consequence of the imperfect correlation between parents and offspring, and that lack of perfect correlation was a necessary requirement for Darwin, or else there would be no intergenerational variation and no natural selection. The

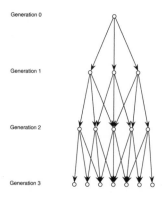

5.13 Figure 5.4 redrawn to allow for regression.

representation of the theory that I gave earlier (Figure 5.4) would better be drawn to incorporate regression, as in Figure 5.13. This differs from Figure 5.4 in that it recognizes that observed stature is not entirely heritable, but consists of two components, of which the transitory component is not heritable. This is then consistent with the population dispersion being in an approximate evolutionary equilibrium, with the movement from the population center toward the extremes being balanced by the movement back, due to the fact that much of the variation carrying toward the extreme is transient excursions from the much more populous middle. The problem Galton had identified was not a problem after all, but was instead due to a statistical effect that no one had identified before. Population equilibrium and intergenerational variability were not in conflict.

Consequences

The consequences of Galton's work on Darwin's problem were immense. He did not play an important role in the acceptance of Darwin's theory—he addressed a problem no one else seems to have fully realized was there and showed that, properly understood, there was no problem—but the methods he developed were extremely important in early twentieth-century biology. He introduced the correlation coefficient and simple variance component models, in effect finding by statistical methods alone some of the results that would flow from the rediscovery of Mendel's work in 1900: the degrees of the quantitative contribution of parents to offspring, and the fact that the relationship among brothers was closer than that between parents and offspring. In 1918, Ronald A. Fisher, in a mathematical tour de force, extended the variance calculations to correlations and partial correlations of relatives under Mendelian assortment to all manner of relationships, and much of modern quantitative genetics was born.[15]

The influence was not only in biology. The variance components idea became key to much quantitative and educational psychology. And Galton's idea of separating permanent and transient effects was at the heart of the model that economist Milton Friedman proposed in 1957 in his *Theory of the*

Consumption Function, for which he won the 1976 Nobel Prize.[16] Friedman argued that individual consumption depended primarily upon the permanent component of an individual's income, and that the individual's consumption was relatively insensitive to transitory increases such as John Maynard Keynes had proposed ("stimulus packages"), and he concluded that economic policy based upon the contrary assumption of a lasting effect to transitory government expenditures was misguided.

Multivariate Analysis and Bayesian Inference

Historians have overlooked one further and arguably even more influential aspect of Galton's discovery. Before Galton's work in 1885, there was no machinery for doing a true multivariate analysis. Earlier workers had considered multidimensional statistical distributions; Figure 5.14 shows some examples of early two-dimensional error distributions, such as those that arose in studies of target shooting, and Figure 5.15 shows early formulas for densities involving more than one variable. Also, formulations involving analyses with more than one unknown go back to the first publication of least squares in 1805 and before.

But before 1885, no one considered slicing the continuous bivariate distribution up, finding for a density of, say, X and

Y the conditional distributions of X given Y and of Y given X: the conditional means and the conditional variances in the normal case. It was a simple mathematical step, but apparently no one was motivated to complete it before Galton was drawn by the study of inheritance to considering the more general problem of bivariate relationships under different conditions.

While Galton was waiting for his 1889 book to make its way through the publication process, he realized that if X and Y had equal standard deviations, the regression slopes for Y on X and X on Y were equal, and their common value could be used as a measure of association: the correlation coefficient was born.[17] Within a few years, Francis Edgeworth, G. Udny Yule, and Pearson had taken the idea to higher

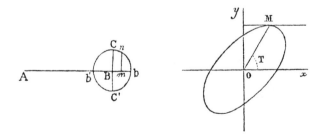

5.14 Bivariate density contours from *(left)* Robert Adrain in 1808, *(right)* Auguste Bravais in 1846, and *(next page)* Isidore Didion in 1858. *(Adrain 1808; Bravais 1846; Didion 1858)*

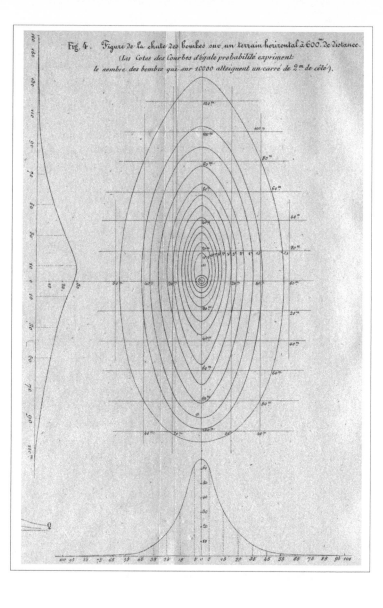

Fig. 4 . Figure de la chute des bombes sur un terrain horizontal à 600.ᵐ de distance.

(Les Cotes des Courbes d'égale probabilité expriment:
le nombre des bombes qui sur 10000 atteignent un carré de 2.ᵐ de côté).

5.15 Published
formulae for
multivariate
normal
densities, from
(above)
Lagrange in 1776
and *(below)*
Laplace in 1812.
*(Lagrange 1776;
Laplace 1812)*

Soit maintenant $x = \xi \sqrt{n}$, $y = \Psi \sqrt{n}$ $z = \zeta \sqrt{n}$ &c.,
& $\frac{a}{n} = A, \frac{\beta}{n} = B, \frac{\gamma}{n} = C$ &c. on aura $\xi + \Psi + \zeta + $ &c. $= o$
& $A + B + C + $ &c. $= 1$; donc,

$$P = \frac{1}{\left(\pi n\right)^{\frac{m-1}{2}} \sqrt{\left(ABC\ldots\right)}} \ \&$$

$$V = e^{-\frac{1}{2}\left(\frac{\xi^2}{A} + \frac{\Psi^2}{B} + \frac{\zeta^2}{C} + \&c.\right)}$$

Or, comme l'incrément où la différence des quantités
x, y, z &c. eſt $= 1$, la différence des variables ξ, Ψ, ζ &c.
fera $= \frac{1}{\sqrt{n}}$ & , par conſéquent, infiniment petite ; de ſorte
que, ſi on appelle cette différence $d\,\theta$, on aura

$$P = \frac{d\,\theta^{m-1}}{\sqrt{\left(\pi^{m-1} ABC\ldots\right)}}$$

Donc $-\frac{1}{2}\left(\frac{\xi^2}{A} + \frac{\Psi^2}{B} + \frac{\zeta^2}{C} + \&c.\right)$

$$P\,v = \frac{e}{\sqrt{\left(\pi^{m-1} A.A.C\ldots\right)}} \qquad d\,\theta^{m-1}$$

$$E = S.m^{(i)2}.S.n^{(i)2} - (S.m^{(i)}n^{(i)})^2;$$

la double intégrale précédente devient

$$c^{-\frac{k}{4k''a^2.E}.[l^2.S.n^{(i)2} - 2ll'.S.m^{(i)}n^{(i)} + l'^2.S.m^{(i)2}]}$$

$$\times \iint \frac{dt.dt'}{4\pi^2.a^2}.c^{-\frac{k'l^2}{k}.S.m^{(i)2} - \frac{k'l'^2.E}{k.S.m^{(i)2}}}.$$

En prenant les intégrales dans les limites infinies positives et négatives, comme celles relatives à $a\varpi$ et $a\varpi'$, on aura

$$\frac{1}{\frac{4k^2\pi}{k}.a^2\sqrt{E}}.c^{-\frac{k}{4k''a^2}.\frac{l^2.S.n^{(i)2} - 2ll'.S.m^{(i)}n^{(i)} + l'^2.S.m^{(i)2}}{E}}. \qquad (o)$$

Il faut maintenant, pour avoir la probabilité que les valeurs de l
et de l' seront comprises dans des limites données, multiplier
cette quantité par $dl.dl'$, et l'intégrer ensuite dans ces limites. En
nommant X cette quantité, la probabilité dont il s'agit sera donc

dimensions, with measures of association related to partial correlations and multidimensional least squares and principal components of variance.[18] Statistics had jumped off the two-dimensional page of tables of numbers, emerging as a technology for handling problems of great complexity.

Bayesian Inference

This new discovery had a striking implication for inference. Fundamentally, inference is the making of conditional statements with data in hand, often based upon formulations outlined before the data are addressed. Bayesian inference is a prime example: In the simplest form, it amounts to specifying a prior probability distribution $p(\theta)$ for an unknown value θ of interest to the statistician, and also a probability distribution for the data X given θ, namely, the likelihood function $L(\theta) = p(x|\theta)$, and then finding the bivariate probability distribution of (X, θ) and from it the conditional probability distribution of θ given $X = x$, the posterior distribution $p(\theta|x)$. At least that is how we do it now. Those simple steps were not possible before 1885. Galton's "look backward" to find the distribution of mid-parent stature given the adult child's stature, or of one brother given that of the other, was a true Bayesian computation and seems to be the first ever made in this form.

Of course, inverse probability of one sort or another has a long history,[19] at least dating back to Thomas Bayes (published 1764)[20] and Laplace (published 1774).[21] But neither these two nor anyone else in the intervening years followed the modern protocol. None of them worked with conditional distributions for continuously varying quantities, and all used the essential equivalent of assuming flat (uniform) prior distributions. Bayes considered only the case of inference for a binomial probability θ (the probability of success in a single trial) for n independent trials, where he supposed the only information available was that a success had occurred (in our terminology) X times and failed $n-X$ times. There was, strictly speaking, no "prior," but he justified his analysis by saying it was equivalent to believing that, absent any empirical evidence, all values of X were equally likely. That is, $\text{Prob}(X=k) = 1/(n+1)$ for all values of $k=0, \ldots, n$. That was consistent with a uniform prior for θ and allowed him to reason to a correct conclusion without recourse to Galton's technical apparatus, but only in the narrow situation of the binomial.[22] Laplace proceeded more generally with the bald assumption (he termed it a "principe") that $p(\theta|x)$ must be proportional to $p(x|\theta)$, also consistent with a uniform prior distribution. Laplace's method, while it did not lead him astray in simple problems, did lead him to serious errors in higher dimensions, although he may not have realized that.[23]

Throughout the nineteenth century, Bayes was generally ignored, and most people followed Laplace uncritically.

Most people consider Bayesian inference the ideal form of inference. It provides exactly the answer scientists seek: a full description of the uncertainty about the goal of an investigation in light of the data at hand. And many people believe that this, like most ideals, is usually unattainable in practice, since the ingredients—particularly the specification of the prior distribution—are often not clearly at hand. After 1885, the mathematical apparatus was available for more general specifications, but difficulties remained. From the 1920s on, Harold Jeffreys argued for the use of what some called reference priors, representations of prior uncertainty that were not sensitive to choice of measurement scale and plausibly (at least to some people) reflected a version of lack of information. In the 1950s, Bruno de Finetti and Jimmie Savage would espouse a personalistic Bayesian inference, in which each statistician would seek an honest assessment of their own beliefs as a prior, even if different individuals thereby came to different conclusions. More recently, others have urged an "objective" Bayes approach, in which a reference prior is again adopted as representing a lack of prior information, with statisticians seeking comfort in the knowledge that any other approach that was not based on strong prior information would reach at least qualitatively similar conclusions. The problems become greater with high dimensions, where the force of assumptions that seem natural in one or two dimensions can be hard to discern and may bring assent despite having very strong and unforeseen effects upon conclusions.

Shrinkage Estimation

The introduction of multivariate analysis was the major achievement of Galton's work. The explanation of the so-called regression paradox—that tall parents may expect to have less tall children, and tall children tend to have less tall parents—was not only less important, it was less successful. Mistakes based upon a misunderstanding of regression have been and remain ubiquitous.

In 1933, the Northwestern economist Horace Secrist published *The Triumph of Mediocrity in Business*, a book founded entirely on that statistical blunder. He observed, for example, that if you identify the 25% of department stores with the highest profit ratio in 1920 and follow their average performance to 1930, it tends unremittingly toward the industry average, toward mediocrity. Even though Secrist knew about regression, he did not understand it, writing, "The tendency to mediocrity in business is more than a statistical result. It is expressive of prevailing behavior relations."[24] He was blissfully unaware that if he had chosen his top 25% based upon 1930 profits, the effect would have been reversed, with a steady move over the period of 1920–1930 *away* from mediocrity. He repeated this blunder separately for dozens of other economic sectors throughout his 468-page book.

In the 1950s, Charles Stein exposed another, related paradox. Suppose that we have a set of independent measurements X_i, $i = 1, \ldots, k$, each an estimate of a separate mean μ_i. The μs may be entirely unrelated, but we suppose for simplicity that each X_i has a normal $(\mu_i, 1)$ distribution. The X_is could be scaled exam scores for k different individuals, or scaled estimates of profits for k firms in different industries. At the time, it was taken as too obvious to require proof that one should estimate each μ_i by the corresponding X_i. Stein showed that this was false if one took as an overall goal the minimization of the sum of the squares of the errors of estimation. In particular, a better set of estimates was to be found by "shrinking" all of the X_i toward zero by an amount that depended upon the X_is only, for example, by using $\left(1 - \dfrac{k}{S^2}\right) X_i$, where $S^2 = \sum_{i=1}^{k} X_i^2$.

Stein's paradox can be interpreted as a version of regression.[25] Consider (μ_i, X_i) as k pairs of observations. Consider as possible estimates all simple linear functions of the X_is; that is, any estimate of the form bX_i. The "obvious" estimate is just this with $b = 1$. But with the minimum sum of squared errors as an objective, if we had the pairs (μ_i, X_i) as data, the best choice of b would be the least squares estimate for the regression of μ_i on X_i. This would be $b = \dfrac{\sum \mu_i X_i}{\sum X_i^2}$. But we do not know the μ_i; indeed, determining them is the point of the analysis. Still, we can estimate the numerator of b. It is an elementary exercise to show that then $E(\mu_i X_i) = E(X_i^2 - 1)$, so replace $\sum \mu_i X_i$ with $\sum (X_i^2 - 1)$ and simply use $b = \dfrac{\sum \left(X_i^2 - 1 \right)}{\sum X_i^2} = \left(1 - \dfrac{k}{S^2} \right)$. Stein proved that this gives (under the assumptions used here) a smaller expected sum of squared errors than the "obvious" estimates no matter what the μ_i are, as long as k is not too small ($k \geq 4$ will do for the estimate given here). Galton would not have been surprised: the "obvious" estimate X_i falls on the line $E(X_i | \mu_i) = \mu_i$, which he would have recognized as the wrong regression line, corresponding to X on μ, not μ on X.

Causal Inference

Today, the statement "Correlation does not imply causation" enjoys universal assent among statisticians. Versions of it even predate the 1888 invention of the correlation coefficient: The philosopher George Berkeley wrote in 1710,

"The Connexion of Ideas does not imply the Relation of Cause and Effect, but only a Mark or Sign of the Thing signified."[26] The modern technical versions seem to date from the late 1890s. In one investigation, Pearson found to his surprise that while the length and width of male skulls were essentially uncorrelated, and the same was true for female skulls, if you took a mixture of male and female skulls, the situation changed: for the combined group, those same measures became markedly positively correlated due to the difference in group means, with both means being different; in his case the male skulls were larger on average in both dimensions. Think of (in an extreme case) the combined group plotted as two disjoint circular clusters, separately showing no relation, but together showing the relation dictated by their centers.

Pearson called this "spurious correlation," writing,

This correlation may properly be called spurious, yet as it is almost impossible to guarantee the absolute homogeneity of any community, our results for correlation are always liable to an error, the amount of which cannot be foretold. To those who persist in looking upon all correlation as cause and effect, the fact that correlation can be produced between two quite uncorrelated characters A and B by taking an artificial mixture of two closely allied races, must come rather as a shock.[27]

The general acceptance of this problem has long coexisted with a strong wish to accept the opposite, to believe that a finding of correlation would indeed support to some degree the inference of a causal relationship. Some of this, of course, is self-delusion, as when a scientist with a strong prior belief in a causal relationship makes incautious statements about his or her own results upon finding correlation. But over the years a collection of statistical techniques has been developed to enable causal inference by potentially permitting statements such as "correlation does imply causation if these assumptions can be agreed upon," followed by a list of conditions that vary with the approach taken.

Some of these lists have been philosophical rather than mathematical. In 1965, Austin Bradford Hill gave a list of seven general statements that he thought necessary for causal inference in epidemiology.[28] These were all sensible, with no attempt at rigorous definitions, using terms like *strength* and *consistency of association, plausibility* and *coherence of the relationship.* One of these seven he called "temporality," essentially stating that the purported cause must precede the effect. But while that seems reasonable in biology or physics, it was not so clear in social science. Simon Newcomb gave this counterexample in a text on political economy, 80 years before Hill's list was published:

This way [that is, assuming temporality] of looking at economic phenomena is so natural that some illustrations of its dangers may be adduced. Let us suppose an investigator seeking to learn the relation between quinine and the public health by statistical observation. He might reason thus: "If quinine conduces to the cure of intermittent fever, then where people take most quinine they will have least intermittent fever, and each new importation of quinine will be followed by an improvement in the public health. But looking at the facts of the case, we find them to be just the reverse of this. In the low lands along the lower part of the Mississippi valley and among the swamps of the Gulf States people take more quinine than anywhere else in the country. Yet, far from being more healthy, they suffer from intermittent fever more than any other people. Not only so, but we find that the large importations of quinine which take place annually in the summer are regularly followed in the autumn by an increase in the frequency of intermittent fevers."[29]

Our ability to anticipate does complicate matters.

Other, more rigorous approaches involve assuming structure in the interdependence of the data, such as that some partial correlations are zero, or that some variables are conditionally independent given some other variables, or by introducing "structural equations" that reflect the assumed causality. In 1917, Sewall Wright discovered that, by constructing a directed graph among the different variables with each direction of dependence indicated by an arrow, he could easily calculate pairwise correlations as long as there were no cycles in the graph (see Figure 5.16).[30] He later called this "path analysis."[31] His initial work was based on the structure of Mendelian inheritance and was mathematical rather than causal in nature, but his later work introduced causal reasoning in some examples. His very first application in 1917 can be used to address to the problem Pearson had discussed. Let L=skull length, W=skull width, and S=sex (M or F). Then his method led to this relationship among covariances for the combined skulls:

$$Cov(L,W) = E\{Cov(L,W|S)\} + Cov(E\{L|S\}, E\{W|S\})$$

For Pearson, the first term on the right would be near zero, and the relationship between the two subgroups means (the second term) would dominate.

Wright's approach foreshadowed much later work, including causal models for acyclic graphical models and economists' structural equation models. Much of that modern work has been conducted in this vein, drawing rigorous conclusions from strong assumptions, with the strong caveat that the assumptions are not usually so demonstrably true as in the case of Mendelian inheritance.

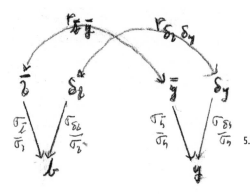

5.16 Sewall Wright's first path analysis of 1917, as reconstructed by Wright in personal correspondence with the author in April 1975. *(Wright 1975)*

The Rule of Three: R.I.P.

By the end of the nineteenth century, the Rule of Three had been consigned to the dustbin of the history of mathematics. Today the rule by that name is essentially unknown to mathematics students and teachers alike, and that name has occasionally been employed for other, unrelated uses, none of which has acquired much of a following. The often quoted question, "Do Jesuits die in threes?" (inspired by the fact that random events can appear to cluster by chance alone) is as close as most statisticians will come to the term *Rule of Three* these days. Only Pearson's repetition of Darwin's 1855 statement on the cover page of the *Annals of Eugenics* survived, even through a later name change to the *Annals of Human Genetics* in

1954, finally to die when the journal's cover was completely redesigned in 1994. The death of the rule went unmourned in a world that no longer understood the reference. Its disappearance is all to the good. But it was not Galton who was responsible; it was the growth and development of mathematics that rendered it such a trivial part of algebra that it no longer merited a name. It once was a named part of every mathematics course and a requirement of the British civil service exam. But even if the name is dead, the statistical misuse of the idea persists in all-too-common unthinking, naive extrapolation.

Even in the heyday of the Rule of Three, the public took a dim view of it, and it probably drove as many school children from mathematics as poorly taught trigonometry and calculus do today. In 1850 John Herschel had already acknowledged the limits to its application in a book review, even without knowing of Galton's insight: "The Rule of Three has ceased to be the sheet anchor of the political arithmetician, nor is a problem resolved by making arbitrary and purely gratuitous assumptions to facilitate its reduction under the domain of that time-honoured canon."[32]

In 1859, a play by Francis Talfourd entitled "The Rule of Three" enjoyed a brief run in London.[33] It was a one-act comedy in which a man, Thistleburr, suspected his lovely wife, Margaret, of being susceptible to the charms of another man, and Thistleburr plotted to undermine what was in fact a nonexistent relationship. His machinations almost cost him his wife, but all ended happily, and the play closed with Margaret reciting a poem to him that reflected the low confidence even then in the Rule of Three as a guide to life:

> You will do well—fling artifice aside;
> A woman's honour is her heart's best guide.
> Trusty while trusted—ne'er to doubt subject it,
> Or 'tis half gone ere only half suspected.
> What fills the vacant place? Ah, who can tell?
> Distrust oft makes the traitor it would quell,
> Nor can the sum of married life e'er be
> Worked out or proved by any Rule of Three.

CHAPTER 6

· DESIGN ·

Experimental Planning and the Role of Randomization

THE SIXTH PILLAR IS DESIGN, AS IN DESIGN OF EXPERIMENTS, BUT interpreted broadly to include the planning of observation generally and the implications for analysis of decisions and actions taken in planning. Design includes the planning of active experimentation, the determination of the size of a study, the framing of questions, and the assignment of treatments; it also includes field trials and sample surveys, quality monitoring and clinical trials, and the evaluation of policy and strategy in experimental science. In all these cases, the planning is guided by the anticipated analyses. Design can even play a crucial role in passive observational science, where there is little or no control over the data generation—any observational study will come into sharper focus by asking, If you had the ability to generate data to address the main question at hand, what data would you seek? As such, design can discipline our thinking in any statistical problem.

Some examples of design are ancient. In the Old Testament's book of Daniel, Daniel balked at eating the rich diet of meat and wine offered to him by King Nebuchadnezzar, preferring a kosher diet of pulse and water. The king's representative accepted Daniel's proposal of what was essentially a clinical trial: For 10 days, Daniel and his three companions would eat only pulse and drink only water, after which their health would be compared to that of another group that consumed only the king's rich diet. Their health was judged by appearance, and Daniel's group won.[1]

The Arabic medical scientist Avicenna (Ibn Sina) discussed planned medical trials in his *Canon of Medicine*, written around 1000 CE. Avicenna's book was a leading medical treatise for six centuries, and in the second volume he listed seven rules for medical experimentation. The rules were cast in terms of the ancient idea that attributed the action of a drug to any of four primary qualities dating from Aristotle (hot, cold, wet, and dry), and they were translated by Alistair C. Crombie as follows:

1. The drug must be free from any extraneous, accidental quality; for example, we must not test the effect of water when it is heated, but wait until it has cooled down.

2. The experimentation must be done with a simple and not a composite disease, for in the second case it would be impossible to infer from the cure what was the curing cause in the drug.

3. The drug must be tested with two contrary types of disease, because sometimes a drug cured one disease by its essential qualities and another by its accidental ones. It could not be inferred, simply from the fact that it cured a certain type of disease, that a drug necessarily had a given quality.

4. The quality of the drug must correspond to the strength of the disease. For example, there were some drugs whose "heat" was less that the "coldness" of certain diseases, so that they would have no effect on them. The experiment should therefore be done first with a weaker type of disease, then with diseases of gradually increasing strength.

5. The time of action must be observed, so that essence and accident are not confused. For example, heated water might temporarily have a heating effect because of an acquired extraneous accident, but after a time it would return to its cold nature.

6. The effect of the drug must be seen to occur constantly or in many cases, for if this did not happen it was an accidental effect.

7. The experimentation must be done with the human body, for testing a drug on a lion or a horse might not prove anything about its effect on man.[2]

With a modern eye, we can read these rules as stressing the need for controls and replication, the danger of confounding effects, and the wisdom of observing the effects for many differing factor levels. One might even look at these rules as an early articulation of causal reasoning in general. Has anything changed since Avicenna? Or, for that matter, since Aristotle? Well, the mouse has replaced the lion as the laboratory animal of choice. But look again at Avicenna's second rule: he says, essentially, to experiment with but one factor at a time. For a more modern version, look at William Stanley Jevons, writing in his *Principles of Science* in 1874:

One of the most requisite precautions in experimentation is to vary only one circumstance at a time, and to maintain all other circumstances rigidly unchanged.[3]

Now read Ronald A. Fisher, writing in 1926:

No aphorism is more frequently repeated in connection with field trials, than that we must ask Nature few questions, or, ideally, one question, at a time. The writer [Fisher] is convinced that this view is wholly mistaken. Nature, he suggests, will best respond to a logical and carefully thought out questionnaire; indeed, if we ask her a single question, she will often refuse to answer until some other topic has been discussed.[4]

Additive Models

Fisher was consigning a large part of two thousand years of experimental philosophy and practice to the dustbin, and he was doing it with a statistical argument of great ingenuity. Fisher's multifactor designs, born of his experience with agricultural research at Rothamsted Experimental Station, constituted a sea change in procedure. He would simultaneously vary seed, fertilizer, and other factors in regular agricultural plots, planting in arrays such as Latin or Greco-Latin squares and answering questions about all these factors in a single planting season, in a single planted array.

Agricultural experimenters had been trying various arrangements for some years. Arthur Young argued in 1770 for careful comparison between methods employed on the same field at the same time, for example, comparing broadcasting seed to drill planting, to avoid being misled by other changing factors.[5] But Young had little to say about just how to divide the field, other than "equally." By the time Fisher arrived at Rothamsted in 1919, researchers there were employing checkerboard and sandwich designs, intended to place two different treatments next to each other so that soil differences would be minimized. This could be said to be an attempt at blocking, although the experiments lacked an analysis that could allow for that effect. The combination of analysis and experimental logic that Fisher provided was radically new. He not only recognized that the statistical variation in agriculture made new methods necessary, but he also saw that if economically efficient experimentation were to be carried out, thinking about the variation also pointed toward the solution. While the most impressive uses of his ideas were in his complex designs, incorporating hierarchical structure and allowing for the estimation of interactions, the gain was already there at a much simpler level.

Consider an additive model for crop yield in an experimental field. The field is divided into a number (say, $I \times J$) of plots and a different combination of treatments is assigned to each. We could express the model algebraically with Y_{ij} as the crop yield for plot (i,j), and suppose the yield was a simple sum of an overall average, with separate effects for the treatments, and a random variation for each plot: For $i = 1, \ldots, I$ and $j = 1, \ldots, J$, let $Y_{ij} = \mu + \alpha_i + \beta_j + \varepsilon_{ij}$, where μ represents the mean yield for the whole field, α_i the effect due to, say, seed variety i, β_j the effect due to, say, fertilizer level j, and ε_{ij} the random variation for plot (i,j) due to uncontrolled factors.

In 1885, Francis Edgeworth had given a verbal description of such an additive model that was not so mathematical but had an eloquent clarity that the formal model does not always convey. "The site of a city," Edgeworth wrote,

> consists of several terraces, produced it may be by the gentler geological agencies. The terraces lie parallel to each other, east and west. They are intersected perpendicularly by ridges which have been produced by igneous displacement. We might suppose the volcanic agency to travel at a uniform rate west to east, producing each year a ridge of the same breadth. It is not known prior to observation whether the displacement of one year resembles (in any, or all, the terraces) the displacements in the proximate years. Nor is it known whether the displacement in one terrace is apt to be identical with that in the neighbouring terraces. Upon the ground thus intersected and escarped are built miscellaneous houses. The height of each housetop above the level of the sea is ascertainable barometrically or otherwise. The mean height above the sea of the housetops for each acre is registered.[6]

Here Y_{ij} could be the mean height of houses above sea level in acre (i,j), μ the mean house height above sea level for the entire city, α_i the effect of ground displacement on terrace i, β_j the effect of ground displacement on ridge j, and ε_{ij} the random variation of house height around the acre mean in acre (i,j).

The point is that this model incorporates three sources of variation in a structured way that permits them to be separately dealt with, even if no two plots have the same combination of treatments. By incorporating all this in one model, there is a huge bonus: If one approaches the data ignoring one factor (for example, fertilizer or ridges), the variation due to the omitted factor could dwarf the variation due to the other factor and uncontrolled factors, thus making detection or estimation of the other factor (for example, variety or terraces) impossible. But if both were included (in some applications, Fisher would call this blocking), the effect of both would jump out through the row or column means and their variation and be clearly identifiable. In even a basic additive effects example, the result could be striking; in more complex situations, it could be heroic.

To give one example that shows clearly what could be missed, consider the famous set of data compiled in the 1890s with great effort from massive volumes of Prussian state statistics by Ladislaus von Bortkiewicz and included in his short tract, *Das Gesetz der kleinen Zahlen* (The law of small numbers) in 1898.[7] The data give the numbers of Prussian

	75	76	77	78	79	80	81	82	83	84	85	86	87	88	89	90	91	92	93	94
G	—	2	2	1	—	—	1	1	—	3	—	2	1	—	—	1	—	1	—	1
I	—	—	—	2	—	3	—	2	—	—	—	1	1	1	—	2	—	3	1	—
II	—	—	—	2	—	2	—	—	1	1	—	—	2	1	1	—	—	2	—	—
III	—	—	—	1	1	1	2	—	2	—	—	—	1	—	1	2	1	—	—	—
IV	—	1	—	1	1	1	1	—	—	—	—	1	—	—	—	—	1	1	—	—
V	—	—	—	—	2	1	—	—	1	—	—	1	—	1	1	1	1	1	1	—
VI	—	—	1	—	2	—	—	1	2	—	1	1	3	1	1	1	—	3	—	—
VII	1	—	1	—	—	—	1	—	1	1	—	—	2	—	—	2	1	—	2	—
VIII	1	—	—	—	1	—	—	1	—	—	—	—	1	—	—	—	1	1	—	1
IX	—	—	—	—	—	2	1	1	1	—	2	1	1	—	1	2	—	1	—	—
X	—	—	1	1	—	1	—	2	—	2	—	—	—	—	2	1	3	—	1	1
XI	—	—	—	—	2	4	—	1	3	—	1	1	1	1	2	1	3	1	3	1
XIV	1	1	2	1	1	3	—	4	—	1	—	3	2	1	—	2	1	1	—	—
XV	—	1	—	—	—	—	—	1	—	1	1	—	—	—	2	2	—	—	—	—

6.1 Bortkiewicz's data were gathered from the large published Prussian state statistics (three huge volumes each year for this period). He included 14 corps (G being the Guard Corps) over 20 years. *(Bortkiewicz 1898)*

cavalry killed by horse kicks in 14 cavalry corps over a 20-year period (see Figure 6.1). Bortkiewicz wanted to demonstrate that the great variability in such small and unpredictable numbers could mask real effects, and he showed that the 280 numbers viewed together were well fit as a set of identically distributed Poisson variables. And indeed they are. But Bortkiewicz lacked the technology of additive models that, if applied here (using a generalized linear model with Poisson variation), clearly shows not only corps-to-corps variation, but also year-to-year variability. The corps and year variations were not large, but the additive model allowed them to be captured by 14 plus 20 separate effects. With 240 separate observations, the analysis could detect these effects, just as the eye could see the ridge and terrace variations in Edgeworth's city by looking only at the tops of houses, notwithstanding the house variation, by looking along the ridges and along the terraces. Bortkiewicz seems to have expected that the corps-to-corps variation was there but masked by random variation; after all, the corps were of different sizes. The yearly variation might have come as a surprise.

Randomization

David Cox has described three roles for randomization in Statistics: "As a device for eliminating biases, for example from unobserved explanatory variables and selection effects; as a basis for estimating standard errors; and as a foundation for formally exact significance tests."[8] The first of these has the widest public appreciation, even finding its way into popular culture. In a July 1977 issue of the comic book *Master of Kung Fu*, the master selects a musical album at random under the thought balloon, "I select one from many by chance. It is refreshing to be free of bias, even though ignorance is the sole liberator." But the other two roles (which are related to each other) are subtler and of greater statistical importance. It is through them that randomization has become fundamental to inference in several ways, particularly in matters of design, and even in some cases by actually defining the objects of inference.

In the late nineteenth century, Charles S. Peirce recognized that the very fact that a sample was random was what made inference possible; he even defined induction to be "reasoning from a sample taken at random to the whole lot sampled."[9] In the early 1880s, Peirce taught logic and did research in experimental psychology at the new Johns Hopkins University. It could be argued that experimental psychology as a field was created by experimental design. In the early work done by Gustav Fechner around 1860 on stimulus and sensation using the method of lifted weights, it was the plan of the experiment that defined and gave meaning to the theoretical goal. The experiment consisted of the experimenter or an assistant lifting each of two small containers successively. Each container held a base weight B and one (and only one) of the two also contained a differential weight D. The person lifting the containers was to guess, based upon the sensation felt in lifting, which was the heavier: Which was B and which was B + D? The experiment was repeated hundreds, even thousands, of times, with varying B and D, and using different hands and different orders of lifting. Some called this the method of right and wrong cases. The data gathered permitted the experimenter to estimate how the chance of a right guess varied with D, B, and the hand doing the lifting, using what we now call a probit model, which assumed the chance rose from 0.5 when $D=0$ to asymptote at 1.0 when D became large. The speed of the rise was taken as measuring the sensitivity associated with the experimental conditions. Without the experiment, the theory was empty or at least unquantifiable. Similarly, in the 1870s, Hermann Ebbinghaus conducted extensive experimental work on the strength of short-term memory, using an elaborate experimental plan employing nonsense syllables.[10]

In 1884–1885, Peirce worked on an even more delicate question that required a further step in experimental methods. Earlier psychologists had speculated that there was a threshold in difference between two sensations such that if the difference was below the threshold, termed the *just noticeable difference* (jnd), the two stimuli were indistinguishable. Peirce, working with Joseph Jastrow, designed an experiment that showed this speculation was false.[11] They worked with a refined version of the lifted weights experiment in which two weights differed by quite small amounts D, with one only slightly heavier than the other. Peirce and Jastrow showed that as the ratio of the weights approached 1.0, the probability of a correct judgment smoothly approached (but remained detectably different from) $\frac{1}{2}$. There was no indication of a discrete threshold effect as claimed by the jnd theory.

Clearly this was a very delicate experiment; the slightest bias or awareness of the order of presentation of the weights would render the enterprise hopeless. Peirce and Jastrow took and documented enormous precautions to ensure the judgments were made blindly, and they introduced a full and rigorous randomization to the order of presentation of the weights (heavy or light first) using a deck of well-shuffled cards. In addition, throughout their study they had the subject note for each judgment the confidence C that he had in the correctness: $C = 0$ (lowest confidence), 1, 2, 3 (highest). They found confidence increased with larger probability p of a correct guess, approximately as a multiple of $\log(p/(1-p))$, the log odds that they were correct—they had discovered evidence that a person's perception of chance is approximately linear on a log odds scale. The validity of the experiment, the inductive conclusion, depended crucially on the randomization.

In the early twentieth century, Fisher took the subject still further. As I mentioned earlier, he recognized the gains to be had from a combinatorial approach to design in multifactor designs. In the five years from 1925 to 1930, as Fisher expanded the complexity of his designs, he saw that the act of randomization could also validate inferences in these complicated settings. In the simplest situation, the random assignment of treatments and controls within pairs made possible valid inference about treatment effects without any distributional assumptions (beyond independence of different pairs). The randomization distribution itself induced a binomial distribution on the number of cases where the chance that treatment surpasses the control under the null hypothesis of "no difference" is $\frac{1}{2}$.

Think back to John Arbuthnot's data on christenings, classified by sex each year for 82 years (see Chapter 3). Those data most assuredly did not arise from a designed experiment. Considering christenings as a surrogate for birth data, Arbuthnot gave the chance of male births exceeding female births for 82 years in a row under a hypothesis of equal probabilities of the two sexes as 1 in 2^{82}. That test could be criticized as an assessment of birth frequency: Were male births really no more likely to be recorded as christened in the parish records than female births? And even if they were, was the rate of infant mortality before being christened the same for both sexes? In his data it was impossible to address these reservations. We become used to accepting such assumptions for observational studies when there is no other option. But imagine (against all logic) that it would have been possible to randomly assign sex at birth as part of an experiment designed to address the hypothesis of no differential christening of males and females. Then, under that hypothesis, the randomization by human design would guarantee the chance of more males being christened in a year is 0.5 (less half the chance of a tie) and Arbuthnot's $\frac{1}{2}^{82}$ would stand as an assessment of the chance of the data under that hypothesis. We might even think back to Chapter 1 and realize that a better test would aggregate further: out of 938,223 christenings in the 82 years, 484,382 were recorded as male, a number 31.53 standard deviations above what an equal probability of sexes would lead us to expect, with a chance not too far off $\frac{1}{2}^{724}$. Of course, Arbuthnot could not randomize, and for him unequal birth frequency and unequal recording were hopelessly confounded. But when possible, randomization itself could provide the basis for inference.

In multifactor field trials, Fisher's randomized designs realized several goals. The very act of randomization (for example, a random choice among Latin square designs) not only allowed the separation of effects and the estimation of interactions, it also made valid inferences possible in a way that did not lean on an assumption of normality or an assumption of homogeneity of material.[12] Fisher recognized that his tests—the various F-tests—required only spherical symmetry under the null hypothesis to work. Normality and independence imply spherical symmetry, but those are not necessary conditions. The design randomization itself could induce a discretized spherical symmetry, granting approximate validity to procedures that seemed to require much stronger conditions, just as randomization of treatments induced binomial variation with Peirce's lifted weights experiment. This subtle point was not widely grasped; even such a wise statistician as William Sealy Gosset ("Student") insisted to the end of his life in 1937 that systematic field trials (such as the sandwich design ABBABBABBA . . .) would provide better estimates than randomized trials and that both required normality. Fisher admired Gosset, but in his obituary article in 1939 he wrote, "Certainly though [Gosset] practiced it, he did not consistently appreciate the necessity of randomization, or the theoretical impossibility of obtaining systematic designs of which both the real and estimated error shall be less than those given by the same plots when randomized; this special failure was perhaps only a sign of his loyalty to colleagues whose work was in this respect open to criticism."[13]

The widespread adoption of these methods was slow and often only partial. A loose form of approximate randomization had long been practiced. Since about 1100 CE, the coins for the trial of the Pyx had been selected "haphazardly," or at least not with conscious bias. In 1895, the Norwegian statistician Anders Kiaer had promoted a method he called "representative sampling," where a purposeful selection would be made with the goal of creating a sample that was, in a not fully specified sense, a microcosm of the population.[14]

In 1934, Jerzy Neyman read an influential paper to the Royal Statistical Society (it was the last occasion upon which he and Fisher enjoyed a semblance of a collegial relationship).[15] Neyman's paper developed the theory of random sampling as a way of achieving Kiaer's goal rigorously. In the discussion, Fisher was approving of that portion of the paper. He did note that the social science application of randomization for only the selection of a sample was generally different from the use Fisher employed in agriculture, where different treatments were imposed at random upon the experimental units. "This process of experimental randomization could not, unfortunately, be imitated in sociological enquiries. If it could, more than was known would certainly be known about cause and effect in human affairs."[16] Over the next two decades, random sampling in the social sciences (in the noninvasive manner) took flight, often with variations such as a focus on subpopulations (stratified sampling) or as part of a sequential process (for example, "snowball sampling").

There was one area where Fisher's more invasive randomization made major inroads: medical or clinical trials. There, the random assignment of treatments was feasible, as Peirce had done with lifted weights and as Fisher had done at Rothamsted, and Fisher's work attracted the notice of Austin Bradford Hill. With Hill's passionate advocacy, the method made slow but steady progress in a medical establishment that was resistant to change.[17] Today, randomized clinical trials are considered the gold standard of medical experimentation, although there are cases where researchers feel they cannot afford gold.

There remains one area where randomized design is widely practiced, but it is never referred to in those terms, and it is often condemned: lotteries. A lottery introduces randomization into a social process and assigns treatments to individuals who volunteer for possible selection. To some, lotteries are for amusement; to others, they are a tax upon stupidity. But they have a long history and show no sign of abating, and it may be worth noting that they can yield some scientific dividends. One example must suffice.

A French lottery was established in 1757; it was patterned on an earlier Genovese lottery and was much like the modern lotto.[18] It continued until it was abolished in 1836, with a break from 1794 to 1797 during the French Revolution's Terror. At regular occasions, five numbered balls would be drawn without replacement and essentially randomly from a set of 90 balls numbered from 1 to 90. Players could bet by specifying all five (a "quine") or a list of four (a "quaterne"), three (a "terne"), two (an "ambe"), or one (an "extrait"). A player would win if his selected numbers were chosen in any order, in any position in the draw of five. At times, the quine bet was disallowed because of the risk of fraud (as when a bribed agent would sell a ticket after the draw was known), but when allowed, it paid 1,000,000 to 1 (a fair bet would pay about 44,000,000 to 1). The odds were better for the more frequently occurring outcomes: an extrait would pay at 15 to 1 (18 to 1 would be fair), an ambe at 270 to 1, a terne at 5,500 to 1, and a quaterne at 75,000 to 1.

Players would usually place several bets at once. For example, for the betting slip shown in Figure 6.2, the player was betting on the six numbers 3, 6, 10, 19, 80, and 90, with 25 centimes bet on each of the six corresponding extraits, 10 bet on each of the 15 ambes, 5 bet on each of the 20 ternes, 5 bet on each of the 15 quaternes, and 5 bet on each of the six quines, for a total of 5 francs 5 centimes. All payoffs were set by a fixed schedule and guaranteed by the king; there was no parimutuel pool to protect the king as there would be to protect the state in modern versions. The actual drawing on the indicated date was 19, 26, 51, 65, and 87; the ticket would have won only on a single extrait (19), paying $25 \times 15 = 375$ centimes, or 3 francs 75 centimes, for a loss of 1 franc 30 centimes. Had the draw given 2, 6, 19, 73, 80, the ticket shown would have been paid for one terne

6.2 An illustrative sample filled-in lottery ticket from an 1800 instruction manual prepared for the loterie staff, showing how to unambiguously write the numerals and how to record and price a combination ticket, where the customer specified six numbers and bet on all five digit subsets, with differing amounts bet on different possible prizes. (*Loterie An IX*)

(6, 19, 80), three ambes (6 and 19, 6 and 80, and 19 and 80), and three extraits, for a payoff of $5500 \times 5 + 3 \times 270 \times 10 + 3 \times 15 \times 25 = 367$ francs 25 centimes.

In its early years, the lottery contributed to the support of the École Militaire; by 1811, the net proceeds furnished up to 4% of the national budget, more than either postal or customs duties. At the date of peak sales around 1810, tickets were sold at over 1,000 local offices and there were 15 draws a month, taking place in five cities (but Parisians could bet on any of these). During the French Revolution, the draws continued uninterrupted through the executions of Louis XVI and Marie Antoinette, and were only suspended when the Terror reach a level where the players' faith in being paid was vanishing. But it was reinstated a little over two years later, when the new regime needed revenue, and drawings continued unabated through the Napoleonic Wars, until the lottery was abolished in 1836 on moral grounds. This was truly randomization on a grand scale.

Throughout its history, the winning numbers were published widely, and this permits a check of the randomness of the draw. The lottery passes all feasible tests, including such tests of the joint occurrence of numbers where the set of 6,606 draws available can reasonably permit a test. This is not surprising; any noticeable bias could only help the players and hurt the lottery.

One social consequence of the lottery was a rise in the level of mathematical education. The players learned combinatorial analysis to assess the bets, and the lottery furnished examples for many textbooks of the time. And the lottery managers had to educate the huge number of local operators so the sales agents could properly price a multiple bet like that shown (they produced special texts for that purpose, such as that furnishing Figure 6.2).

As another benefit, the lottery was inadvertently executing what is likely the earliest scientifically randomized social survey. In the period after the French Revolution, not only were the winning numbers published, but they also published all the winning bets at the level of the quaterne or higher. These records gave the amount paid out (from which the size of bet could be found) and the location and number of the office where the ticket was sold. These winners were a truly random selection of those placing quaterne bets, and so this survey gives a picture of where in France the interest in the lottery was greatest (Paris, of course, but interest was high across the country), and how the attraction to the lottery changed over time. On that last question, the results show a steady decrease in betting over the last two decades, leading one to believe that only when the profits dropped below the level needed to sustain the large operation did the "moral issue" dominate policy.

CHAPTER 7

· RESIDUAL ·

Scientific Logic, Model Comparison, and
Diagnostic Display

I CALL THE SEVENTH AND FINAL PILLAR RESIDUAL. The name
suggests a part of standard data analyses, and while that is
not entirely misleading, I have a larger and more classical
topic in scientific logic in mind.

John Herschel's father, William Herschel, was the discoverer of Uranus, and John followed his father into astronomy.
But whereas his father's second career was in music, John
Herschel's was in mathematics and scientific philosophy, and
he became one of the most honored and respected scientists
of his generation. He discussed the process of scientific
discovery in a widely read and influential book published
in 1831, *A Preliminary Discourse on the Study of Natural Philosophy.*

Herschel gave particular emphasis to what he called residual phenomena.

> Complicated phenomena, in which several causes concurring, opposing, or quite independent of each other, operate at once, so as to produce a compound effect, may be simplified by subducting the effect of known causes, as well as the nature of the case permits, either by deductive reasoning or by appeal to experience, and thus leaving, as it were, a *residual phenomenon* to be explained. It is by this process, in fact, that science, in its present advanced state, is chiefly promoted. Most of the phenomena which nature presents are very complicated; and when the effects of all known causes are estimated with exactness, and subducted, the residual facts are constantly appearing in the form of phenomena altogether new, and leading to the most important conclusions.[1]

From a historical perspective, he made an unfortunate choice for an example: he attributed to this kind of reasoning the "discovery" of the aether (later called the "luminiferous ether") that was believed to fill outer space and both convey light and be responsible for some anomalies in Newtonian theory. We are still looking for that aether. But the scientific principle was valid and important: we can learn by trying explanations and then seeing what remains to be explained.

Among those influenced by Herschel's book were Charles Darwin, said to have been thus inspired to be a scientist, and John Stuart Mill, who in his book *A System of Logic*, published in 1843, presented Herschel's idea as the most important of the four methods of experimental inquiry. He slightly altered the name Herschel gave, calling it "the Method of Residues," and Mill wrote, "Of all the methods of investigating laws of nature, this is the most fertile in unexpected results; often informing us of sequences in which neither the cause nor the effect were sufficiently conspicuous to attract of themselves the attention of observers."[2]

The idea, then, is classical in outline, but the development of this idea in Statistics has made it into a new and powerful scientific method that has changed the practice of science. The statistical interpretation of this idea, and the associated scientific models, has given it a new disciplinary force. The statistical approach is to describe the process that generates the data by a hypothetical model and examine the deviation of the data from that model either informally (for example by a graphical or tabular display) or by a formal statistical test, comparing the simpler model with a more complicated version (a comparison among two "nested" models, one being a special case of the other).

The earliest instances involved small, focused nested models, where one theory is to be compared to a slightly more complicated version. A prime example of the simplest type is the eighteenth-century study of the shape of the earth discussed in Chapter 1. There, the basic (smaller) model is the earth as a sphere. To test that, we concoct a slightly more complicated model, with the earth as an ellipsoid—a sphere either squashed or elongated at the poles. We subtract the sphere from the earth and see in what direction the residual suggests the departure, if any, would seem to go. But how to do that—what measurement of the earth should be employed for such a test? The route taken in the eighteenth century was to measure a series of short arc lengths along a meridian arc. If A is the length of $1°$ arc at latitude L, then if the earth were a sphere A should be the same at all latitudes, say $A=z$. But if the earth were an ellipsoid, then to a good approximation you should have $A=z+y \sin^2(L)$. If the earth were squashed or elongated at the poles, you would have $y>0$ or $y<0$. So, given a set of arc measures at different latitudes, the question was, Did the residual from fitting the equation with $y=0$ (that is, $A=z$) slope upward or downward or not at all, as $\sin^2(L)$ increased? Were degrees nearer a pole shorter or longer than those near the equator?

Figure 7.1 shows Roger Joseph Boscovich's own plot with the line for the sphere ($A=z$, where z is the average arc length for the data) as XY, and the five data points as a, b, c, d, and e are then the residuals from XY. The horizontal axis is AF,

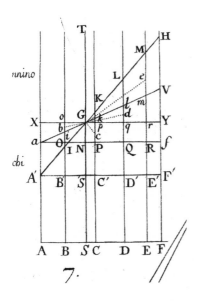

7.1 Boscovich's own plot from a 1770 publication shows the line for the sphere as *XY*, and the five data points as *a*, *b*, *c*, *d*, and *e* are the residuals from *XY*. The line found by Boscovich from his algorithm is *GV*, the slope for the residuals from *XY*. *(Maire and Boscovich 1770)*

giving values of $\sin^2 L$ from $A = 0$ to $F = 1$, the vertical axis is *AX*, giving values of arc length *A*. The line found by Boscovich from his algorithm, that is, the line that goes through the center of gravity *G* of these five points and minimizes the sum of the absolute vertical deviations from the line, is shown as the line *GV*. It gives a positive slope for the residuals from *XY*: the earth is squashed.

This situation was the simplest possible case of a nested pair of models, $A = z$ being a special case of $A = z + y \sin^2(L)$. It is an early example of comparing one regression equation with another where a new "predictor" is added, namely,

$\sin^2(L)$. While Boscovich made no mention of probability (and so gave no estimate of the uncertainty attached to the estimate of y), the basic method he used has been the mainstay of statistical exploration of model adequacy ever since.

These nested models arose naturally in the physical sciences: when the simpler model was made more complicated and then linearized as a local approximation, it tended to add one or more terms to the equation. Measures were taken corresponding to different points of the equation, subject to experimental error, and after 1805 the evolving technique of least squares made the comparison simple. Were the additional terms zero, or were they not? That was the question when Leonhard Euler, Joseph-Louis Lagrange, and Pierre Simon Laplace tried different ways to introduce three-body attraction into the two-body Newtonian model for planetary motion in the 1700s. Their studies were set in motion by an observed residual effect. As the orbits of Saturn and Jupiter became better determined and data over a long span of time came under closer scrutiny, it appeared that over the past few centuries Jupiter had been accelerating and Saturn retarding. If these massive bodies were in unstable orbits, it did not bode well for the solar system. The suspicion was, though, that the changes were only an artifact of a three-body attraction: the sun, Jupiter, and Saturn. Euler and Lagrange made important progress, but it was Laplace who finished the study by finding a way to expand the equations of motion and group the higher order terms in a way that allowed a test of whether the apparent motions were in line with the effects of those terms. His success revealed that the changing speeds observed were but one part of an approximately 900-year periodic change in motion attributed to the nearly 5 to 2 ratio in those two planets' mean motions.[3] A residual analysis had saved the solar system.

These studies set an example that many others followed. By the 1820s, some of the comparisons were done via a test of significance, or at least with a comparison between the unexplained discrepancy and an estimate of the probable error (or p.e.) of the additional coefficients. Laplace's 1825–1827 study of the lunar tide of the atmosphere was of that sort (see Chapter 3). The approach did not give an easy route for comparing non-nested models, where neither model represented the other less a residual. But then, there is still no generally accepted approach to that intrinsically harder philosophical question, where the meaning of "simpler" does not come easily from the statement of the models.

From about 1900 on, when the linear equations of the physical sciences began to be adapted to new uses by the social sciences, it was natural to follow the same route: specify one set of "explanatory" variables, then add a few more linearly and see if they made a significant difference or if their terms were not statistically different from zero. One of the earliest fully developed examples was G. Udny Yule's examination in 1899 of a part of the English welfare system, the "Poor Laws."[4] In one investigation, he looked at the relationship between the poverty level and the amount of welfare relief: Did increasing welfare relief lead to an increase or a decrease in the poverty level? He compared 1871 data and 1881 data and sought to determine the effect a change in the proportion of a municipal district receiving "out-relief" (welfare) had upon the decade's change in "pauperism" (poverty level) in that district (see Figure 7.2). That would have been a simple regression with different districts giving the different data points, but Yule knew that other economic factors changed as well. So he instead reframed the question as one about a residual phenomenon: Correcting for those of the other economic factors for which he had data, what was the relationship? In a long and carefully nuanced analysis, he found a positive relation between these two variables.

Questions of the interpretation of this result and of the difficulty of inferring causation from correlation were as vexed then as they are now (and were carefully noted by Yule), but this marked the beginning of a new era in social science research. The approach faced many complications:

used. Then suppose a characteristic or regression equation to be formed from these data, in the way described in my previous paper, first between the changes in pauperism and changes in proportion of out-relief only. This equation would be of the form—

> change in pauperism
> $=A+B \times$ (change in proportion of out-relief) $\Big\} -$
> where A and B are constants (numbers)   (1)

This equation would suffer from the disadvantage of the possibility of a double interpretation, as mentioned above: the association of the changes of pauperism with changes in proportion of out-relief might be ascribed *either* to a direct action of the latter on the former, *or* to a common association of both with economic and social changes. But now let all the other variables tabulated be brought into the equation, it will then be of the form—

change in pauperism=

$a+b \times$ (change in proportion of out-relief)
$+c \times$ (change in age distribution)
$+d \times \Big]$
$+e \times \Big\}$ changes in other economic, social, and moral factors.
$+f \times \Big]$ (2)

Any double interpretation is now—very largely at all events—excluded. It cannot be argued that the changes in pauperism and out-relief are both due to the changes in age distribution, for that has been separately allowed for in the third term on the right; $b \times$ (change in proportion of out-relief) gives the change due to this factor *when all the others are kept constant*. There is still a certain chance of error depending on the number of factors correlated both with pauperism and with proportion of out-relief which have been omitted, but obviously this chance of error will be much smaller than before.

7.2 Yule's multiple regression equation for change in pauperism.
(Yule 1899)

the linear equation was open to question (although Yule noted he estimated the closest linear approximation to any more complicated relation), and interrelations among the "explanatory" variables could greatly muddy the interpretation. If used carefully, it gave a powerful exploratory technique. Still, except for one important advance, the technique seemed to be limited to linear least squares. That advance was the introduction of parametric models.

One of Ronald A. Fisher's subtlest innovations was the explicit use of parametric models. This innovation can be easily overlooked because he himself never called attention to it. But in Fisher's fundamental 1922 paper introducing a new theoretical mathematical statistics, the word *parameter* is seemingly everywhere, while it is nearly absent in earlier statistical work by Fisher or anyone else.[5] By replacing Karl Pearson's very general unspecified population distribution $f(x)$ with a family of distributions $f(x; \theta)$ that were smooth functions of the possibly multidimensional parameter θ, Fisher gave restrictions and structure to the problems he studied in estimation or testing, permitting mathematical analyses that were impossible before. In retrospect, we can think of the linear models of earlier least squares as special cases of parametric models, but Fisher's extension was much more general and bore unexpected theoretical fruit.

In articles published from 1928 to 1933, Jerzy Neyman and Karl Pearson's son Egon S. Pearson took Fisher's innovation and turned it into an approach to testing hypotheses that from the beginning was designed to test models. The strongest result, widely known as the Neyman-Pearson Lemma,[6] even gave an answer to the comparison of two non-nested models, but only if they were completely specified—there were no unknown parameters to estimate, and it was a direct question: Did the data come from sampling distribution A or from sampling distribution B? Much more general was the generalized likelihood ratio test, where the test was explicitly of the residual type: The test could be viewed as a contest where a parameterized family of distributions was pitted against a much broader family that included the first. Of course, the larger family, because of its greater flexibility, would fit more closely. But would the gain in the additional flexibility be sufficient to justify its use—specifically, was the gain more than chance alone would lead us to expect?

For a simple example, consider a question that Karl Pearson studied in 1900.[7] In a heroic effort to gain a better understanding of chance, Frank Weldon had thrown 12 dice at a time and for each group of 12 counted how many of the dice showed a 5 or a 6. He repeated that experiment a total of 26,306 times. The total number of single die tosses was then $12 \times 26,306 = 315,672$. Pearson's table gave the results in the "Observed Frequency" column.

To give you some idea of the labor involved, I can report that a student in one of my courses, Zac Labby, repeated the experiment a few years ago.[8] He invented a mechanical way of doing it—a box with 12 ordinary dice was shaken, the dice were allowed to settle, and the computer took a picture of the result. The pictures were then computer processed to get the data. A single trial was accomplished in about 20 seconds. The machine ran night and day and took about a week to perform 26,306 trials. Imagine one man doing all that by hand. In one report, Weldon did indicate that his wife had assisted him; one can still wonder about the stress on the marriage.

The point of Weldon's experiment was to see how closely the real world approximated theory. On each trial there were 13 possible outcomes, 0, 1, . . . or 12. If the dice were exactly fair and exactly independently thrown, theory would say that the chance of "5 or 6" was $\frac{1}{3}$ for each of the 12 dice for a single trial, so the chance that k of the 12 were "5 or 6" would be the binomial probability Prob{# "5 or 6"=k} = $\binom{12}{k}\left(\frac{1}{3}\right)^k\left(\frac{2}{3}\right)^{12-k}$. The "Theoretical Frequency" column is a list of these, each times 26,306.

Before (and even during) 1900, this was an extremely difficult question to treat properly. The data gave a point in a 13-dimensional space (Pearson's m' in Figure 7.3), and such structures were not familiar to the statistical world. Was the set of 26,306 trials in the aggregate so far from the theoretical point (Pearson's m) that the binomial model should be discarded? And if that hypothesis were to be discarded—then in favor of what? Looking separately at each of the 13 dimensions (the 13 rows in the table) was

No. of Dice in Cast with 5 or 6 Points.	Observed Frequency, m'.	Theoretical Frequency, m.	Deviation, e.
0	185	203	− 18
1	1149	1217	− 68
2	3265	3345	− 80
3	5475	5576	−101
4	6114	6273	−159
5	5194	5018	+176
6	3067	2927	+140
7	1331	1254	+ 77
8	403	392	+ 11
9	105	87	+ 18
10	14	13	+ 1
11	4	1	+ 3
12	0	0	0
	26306	26306	

7.3 Weldon's dice data, as given by Pearson. *(Pearson 1900)*

seen by the better analysts of the time (for example, Francis Edgeworth and Pearson) as incorrect. But then what to do? Pearson's solution through the derivation of the chi-square test was a truly radical step in 1900—a first appearance of a way to do multiple tests all at once, to include 13 dimensions in one test in a way that compensated not only for the fact that 13 questions were involved, but also for the obvious dependence among the 13. If the set of points was far from theory in one dimension, it would likely be far from theory in others; if no 5s or 6s occurred at all, some other outcomes must occur too frequently. The chi-square test put the simple model to test against a very broad set of other explanations, explanations that included all possible other

Group.	m'.	m.	e.	e^2/m.
0	185	187	− 2	·021,3904
1	1149	1146	+ 3	·007,8534
2	3265	3215	+ 50	·777,6050
3	5475	5465	+ 10	·018,2983
4	6114	6269	−155	3·991,8645
5	5194	5115	+ 79	1·220,1342
6	3067	3043	+ 24	·189,2869
7	1331	1330	+ 1	·000,7519
8	403	424	− 21	1·040,0948
9	105	96	+ 9	·841,8094
10	14	15	− 1	·666,6667
11	4	1	+ 3	9
12	0	0	0	0

7.4 Weldon's dice after Pearson recomputed the theoretical values m using $\theta = 0.3377$. *(Pearson 1900)*

binomial distributions *and* all non-binomial distributions, as long as the 26,306 trials were independent.

Pearson's test found that the data were not consistent with the simple model, that something else was afoot. And the data suggested that "5 or 6" occurred more frequently than one time out of every three dice. The data themselves gave the fraction of 5s or 6s as $106{,}602/315{,}672 = 0.3377$, slightly larger that $\frac{1}{3}$. Pearson tried next to test the general binomial hypothesis, where Prob$\{\#$ "5 or 6"$= k\} = \binom{12}{k}(\theta)^k (1-\theta)^{12-k}$, but without insisting that $\theta = \frac{1}{3}$, and when $\theta = 0.3377$ was used to compute the theoretical values, the fit was improved (see Figure 7.4).

In fact, the data passed the chi-square test for the newly computed column of ms. Based upon his new conception of parametric models, Fisher would show in the early 1920s that Pearson erred here in making no allowance for the fact that he was essentially letting the data choose the theoretic values, but in this case the error was not a serious one and the conclusion withstood Fisher's correction for a lost "degree of freedom."[9]

Pearson guessed that the reason for the larger $\theta = 0.3377$ was that the dice, like many then and in later times, had each pip formed by scooping out a tiny amount of the material of the die at that spot, and the sides with 5 and 6 were then the lightest of the six sides, even if by only a minute amount. Throughout the following century that guess stood, and it seemed convincing to all who heard about it. But when Labby recreated the experiment with the same kind of dice, something surprising occurred.[10] Working via computer counts, he was able to get a count of each of the six possible outcomes for each die, unlike Weldon and his wife who only noted "5 or 6" or "not 5 or 6." Labby found (like Pearson) that the outcomes were not in line with the simple hypothesis; he found a fraction 0.3343 of the dice came up "5 or 6." And he found that the most frequent side to come up was indeed 6. But there was a surprise: the second most frequent face was 1. And this suggested an alternative explanation. The faces 6 and 1 were on opposite sides of each die. Maybe the dice were not cubical; maybe those two sides were closer together than any other pairs of opposite sides. Maybe the dice were like fat square coins, with 1 and 6 as heads and tails, and the other sides being the edges. When Labby went back to the dice with a precision caliper, the results did support that conjecture! The 1–6 axis for his dice was shorter by about 0.2%.

Pearson's test opened new statistical vistas for the practice of the residual method. With Fisher's correction on the question of degrees of freedom, and various extensions to more complicated parametric models, it was now possible to test even some quite complicated hypotheses against an omnibus of much more complicated alternatives, all in one test. In the 1970s, generalized linear models were introduced, incorporating likelihood ideas in a very flexible way that extends to all manner of parameterized counted data, as well as to standard linear models and the analysis of variance, and even extends to the use of tests within nested models across that platform.

The extensions are not limited to parametric models. David Cox had the insight to see that all that was really necessary for the residual tests I discuss was that the additional portion be parameterized: the base model could be nonparametric.[11] That is, in using the notion of comparing a simpler model with a more complicated model of which it is a special case, the "simpler model" did not need to be simple; it could be quite complicated and indeed not even always well specified. What was really needed was that the additional portion be parametric, in order to allow the use of powerful parametric methods to make rigorous tests of the gain in explanatory power. Cox's methods, sometimes called partial likelihood methods, have had a huge impact through the use of Cox regression models in survival data analyses and other applications in medicine.

Diagnostic and Other Plots

The most common appearance in Statistics of the term *residual analysis* is in model diagnostics ("plotting residuals"). Among statisticians it has become common practice after fitting a regression model to plot the "residuals" (=observed dependent variable minus fitted value) to assess the fit and to see what the pattern might suggest for a next stage in modeling. For example, Figure 7.5 shows two residual plots, the first for a regression $S=a+bA+cE$, for a set of data from $n=23$ of the Galapagos Islands,[12] recording number of endemic species S and the area A and elevation E at the highest point for each island, the object being to understand how A and E can

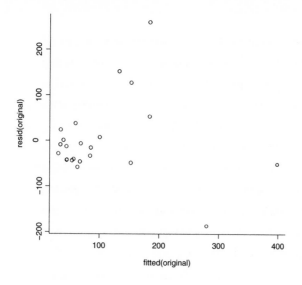

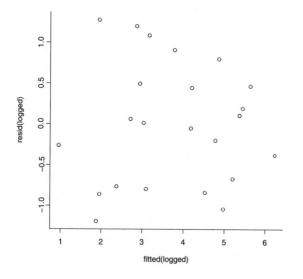

7.5 The residual plots for the Galapagos data *(above)* on the original scale, and *(below)* after transformation to log scales.

affect species diversity S. The 23 residuals from this fitted model are the differences $S - \hat{s}$, where each $\hat{s}$ is the island's value of $a + bA + cE$ when a, b, and c are the least squares estimates. The plot of $S - \hat{s}$ versus $\hat{s}$ suggests a relationship with increasing variation with larger $\hat{s}$. It suggests transforming the variables by taking logarithms, and leads to the much more sensible model $\log S = a + b\log A + c\log E$, whose residual plot is in the second panel. The second model describes a multiplicative relationship $S \propto A^b E^c$. One consequence of the analysis is the realization that, while the Galapagos may have been a wonderful site for Darwin's exploration, they are less wonderful for separating the effects of area and elevation on species diversity. To a rough approximation the islands are volcanic cones, and E is roughly proportional to the square root of A. And indeed, looking at the variables on a log scale shows $\log A$ and $\log E$ are approximately linearly related. To separate the effects other data are needed.

Statistical graphics have a long history. They found interesting application in the 1700s, but they only really flourished in the 1900s, and their use has exploded in the computer age, to the point where, on occasion, a thousand pictures are worth but a single word. If we set aside those graphics that only serve as decoration (and that is a significant portion of their use today), it is only a slight simplification to say that all of the remainder are deployed either

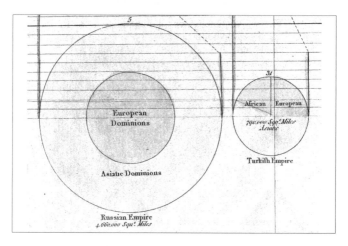

7.6 The first pie chart. *(Playfair 1801)*

as tools of rhetoric or as tools for diagnosis and discovery. Residual plots fall into the latter category, but in fact all diagnostic plots are to a degree residual plots, using the present expansive definition of *residual*. Even a lowly pie chart, when it has any value beyond decoration, is a way of showing a degree of inequality in the various segments, through the chart's departure from the baseline of a pie with equal pieces (see Figure 7.6).

In 1852, William Farr published a different sort of circular plot as part of a study of the cholera epidemics that had swept through England in 1848–1849.[13] He wanted to discover the mechanism responsible for the spread, and he

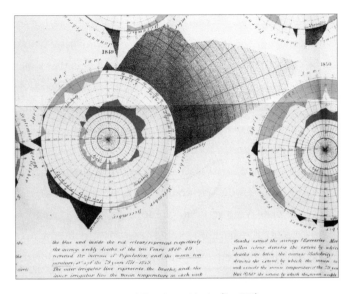

7.7 Farr's circular plot of the 1849 cholera epidemic. *(Farr 1852)*

plotted several variables around a circle for each year's data. Figure 7.7 shows the data for 1849. (The original was printed in color.) The outer circle shows mortality; the circle is the baseline at average annual mortality per week, absent an epidemic, and the total number of deaths was marked for each week's data as the distance from the center for that week. It shows a huge number of deaths during July–September 1849. The below-average mortality in May and November is shown in a lighter shade, dipping below the outer circle. The great cholera epidemic of July–September 1849 seems to jump off the page. Farr suspected the cause was airborne, but the plots did not provide an answer, and only later would he be convinced based on nongraphic arguments that water was the means of the spread of that disease. The inner circle serves as a baseline and displays mean weekly temperature; the circle itself gives the annual average, and the higher temperatures from June to September track the cholera epidemic with a slight lead. All together, there is the apparent relationship between disease and climate, with both parts of the chart displaying residual phenomena.

Farr's plots did have one other important consequence. Florence Nightingale picked up the idea from Farr and employed it to great rhetorical effect in her determined effort to reform the sanitation practices in British field hospitals (see Figure 7.8).

Nightingale had served in the military hospitals during the Crimean War in Scutari (in Turkey, near the Crimea), and she knew that the high mortality there was due not to battle injuries but to diseases like cholera and to insufficient sanitary policy. In contrast, the military hospitals in Britain were able to cope with such problems and had a much better record. She returned to England determined to press publicly for a higher standard for field hospitals. She adopted the form of Farr's plot to show with the large wedges the mortality in Scutari and Kulali, in contrast to the baseline

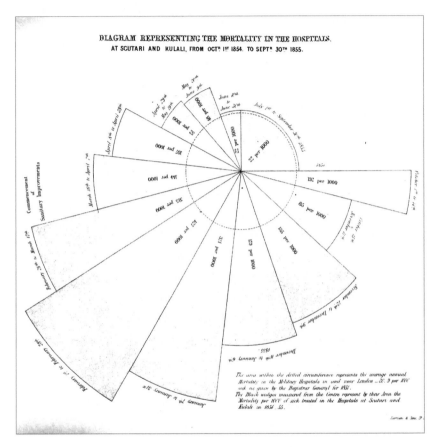

7.8 Nightingale's diagram showing mortality in the field hospitals of the Crimean War. *(Nightingale 1859)*

of the average mortality in military hospitals in and near London, shown by the dotted circle, dramatically smaller than the wedges. Nightingale made one change from Farr, and it was surprising. Whereas Farr showed the mortality figures as distances from the center, Nightingale represented them as the *areas* of the wedges; that is, she plotted the *square roots* of deaths as distances from the center (this would have involved considerable labor to construct at the time, but I have recalculated the heights from the data and she did indeed do as she claimed).

Farr's plot was deceptive in the impression it gave—if you double the mortality under his system, you quadruple the area and the visual effect is exaggerated. Nightingale's plot did away with that and gave a picture that did not mislead. It is ironic that Farr, who plotted in hopes of discovery, produced a misleading picture, while Nightingale, who plotted for rhetorical purposes, did not mislead. Both emphasized residuals from an average.

· CONCLUSION ·

THE SEVEN PILLARS ARE THE PRINCIPAL SUPPORT for statistical wisdom; they do not by themselves constitute wisdom. They all date back at least to the first half of the twentieth century, some to antiquity; they have proved themselves by long use and yet they adapt to new uses as needed. They are the basis of the science of Statistics, the original and still preeminent data science; they can be viewed as an intellectual taxonomy of that science. They can partner well with other information sciences, such as computer science and others with new names that have not yet acquired a full identity. And yet these are still radical ideas, dangerous if misapplied and capable of eliciting an antagonistic response when invading unfamiliar territory. None of them is out of date, but we may still ask if more is needed in the modern age. Should we hew an eighth pillar? And if so, to what end? As a statistical approach to this question, let us review the data, the seven pillars, to see if they suggest an answer.

The first, Aggregation, inherently involves the discarding of information, an act of "creative destruction," in a term used by Joseph Schumpeter to describe a form of economic reorganization, which is another way to view the act. As in such other uses, it must be done on principle, discarding information that does not aid (or may even detract from) the ultimate scientific goal. Even so, it can be accused of rendering invisible the individual characteristics that can from other perspectives be a part of the goal. How can a "personal medical information system" be developed without individual characteristics? In some statistical problems, a notion of a sufficient statistic—a data summary that loses no relevant information—can be employed, yet, in the era of big data, that frequently is not feasible or the assumptions behind it are untenable. Balancing these concerns is a necessary part of supporting statistical wisdom.

The second, Information and its measurement, takes on a different meaning in Statistics from that found in signal processing. It works with aggregation to help recognize how the diminishing rate of gain in information relates to the anticipated use, how this may help plan both the experiment and the form of aggregation. In signal processing, the information passed can remain at a constant rate indefinitely; in Statistics, the rate of accumulation of the information from the signal must decline. The realization that seemingly equivalent blocks of information are not equally valuable in statistical analysis remains paradoxical.

The third, Likelihood, the use of probability to calibrate inferences and to give a scale to the measurement of uncertainty, is both particularly dangerous and particularly valuable. It requires great care and understanding to be employed positively, but the rewards are great as well. The simplest such use is the significance test, where misleading uses have been paraded as if they were evidence to damn the enterprise rather than the particular use. Its growing use over the past century is testimony to the need for a calibrated summary of evidence in favor of or against a proposition. When used poorly the summary can mislead, but that should not blind us to the much greater propensity to mislead with verbal summaries lacking even a nod toward an attempt at calibration with respect to a generally accepted standard. Likelihood not only can provide a measure of our conclusions, it can be a guide to the analysis, to the method of aggregation, and to the rate at which information accrues.

The fourth, Intercomparison, gives us internal standards and a way to judge effects and their significance purely within the data at hand. It is a two-edged sword, for the lack of appeal to an outside standard can remove our conclusions from all relevance. When employed with care and intelligence, it, together with the designs of the sixth pillar, can yield an almost magical route to understanding in some high-dimensional settings.

The fifth pillar, Regression, is exceedingly subtle. It is a principle of relativity for statistical analysis, the idea that asking a question from different standpoints leads not only to unexpected insight but also to a new way of framing analyses. The subtlety is attested to by the late date of discovery, in the 1880s. The idea is not simply the construction of multivariate objects; it is the way they are used, taken apart and reassembled in a genuine multivariate analysis. Inverse probability in rudimentary forms is relatively old, but before the 1880s there was no mechanism to describe inference in general, and Bayesian inference in particular. The earlier attempts can be likened to flight by a glider—at best, it is slowly falling but will give the illusion of flight in limited terrains under ideal circumstances. With the developments in the 1880s, we had powered flight to soar in principle in all circumstances, and to avoid the mishaps or impossibilities that had proved fatal to some earlier explorers. Once more fully developed in the twentieth century, the methods that flowed from this understanding could empower tours to higher altitudes and even to higher dimensions, a trick that more mundane methods of transport have yet to accomplish.

The sixth pillar, Design, also involved great subtleties: the ability to structure models for the exploration of high-dimensional data with the simultaneous consideration of multiple factors, and the creation through randomization of a basis for inference that relied only minimally upon modeling.

The final pillar, Residual, is the logic of comparison of complex models as a route to the exploration of high-dimensional data, and the use of the same scientific logic in graphical analysis. It is here that in the current day we face the greatest need, confronting the questions for which we, after all these centuries, remain least able to provide broad answers. It is here that we may see the potential need for an eighth pillar.

With ever larger data sets come more questions to be answered and more worry that the flexibility inherent in modern computation will exceed our capacity to calibrate, to judge the certainty of our answers. When we can limit attention to a few alternatives or to well-structured parametric models, we are comfortably at home. But in many situations that comfort is missing or illusory. Consider, for example, these three types of problems: (1) formulation of predictions or classifiers with big data—data on many individual cases with many dimensional measures on each case; (2) large multiple-comparison problems; and (3) analyses in cases where the focused questions follow as a final stage of a scientific study that was at least in part exploratory.

In the first of these, we are faced with problems inherent in any exploration in high dimensions. Suppose we are constructing a prediction of some measured response in terms of 20 characteristics—the predictors are in 20-dimensional space, a common event in machine learning. How large is 20-dimensional space? If we divide each predictor's range into quartiles, the 20-dimensional space is divided into 4^{20} different sections. If you have a billion individual cases, on average there will be only one case in every thousand sections. Hardly an empirical base to build upon with confidence! And so any reasonable analysis must (and does, even if only implicitly) make highly restrictive assumptions, perhaps through a low-dimensional parametric model, or at least by assuming that the data are close to some low-dimensional subspace. Under such assumptions, many excellent algorithms have been devised within the area of machine learning. Generally that excellence has the limited support of successful applications in some cases, with little evidence of general applicability. In one case, so-called support vector machines, statistician Grace Wahba has shown they can be viewed as approximating particular Bayesian procedures, thereby adding greatly to our knowledge of how they may be extended by shedding light on why and when they work so well. But the general problem remains quite difficult.

In the second type of problem, multiple comparisons, we are faced with the prospect of making a potentially very large number of tests. In the analysis of variance this could be the comparison of many factors' effects via confidence intervals for a very large number of pairwise comparisons. In genomic studies, thousands of different sites may be put to separate significant tests that are not independent of one another. Probability calibration—confidence intervals or significance tests—which is appropriate when only one pair or one case is available, is not useful if the pair or case was selected as among the more extreme in half a million cases. Even in the 1960s it was known that procedures designed then by John W. Tukey and Henry Scheffé to compensate for such selection by weakening the resulting statements, as with larger confidence intervals, were not the full answer. David Cox saw a part of the difficulty in 1965: "The fact that a probability can be calculated for the simultaneous correctness of a large number of statements does not usually make that probability relevant for the measurement of the uncertainty of one of the statements."[1] Cox was noting that overall corrections (such as by Tukey or Scheffé) do not condition on peculiarities of the data at hand and may be much too conservative. More modern concepts such as false discovery rates are being developed, but the problem remains difficult.

The third type of problem, in which the focused questions arise late in the analysis, is related to the first two but is more general. Even in small data problems the routes taken may be many, so many that from one point of view it becomes effectively a large data quandary. Alfred Marshall realized this back in 1885, writing that "the most reckless and treacherous of all theorists is he who professes to let facts and figures speak for themselves, who keeps in the background the part he has played, perhaps unconsciously, in selecting and grouping them, and in suggesting the argument *post hoc ergo propter hoc*."[2] Andrew Gelman has borrowed an apt term from the title of a 1941 Jorge Luis Borges story to describe the problem: "the garden of forking paths," when a conclusion has been judged reasonably certain after a tortuous journey involving many choices (of data, of direction, of type of question) that are not taken into account in the final assessment of significance.[3] Big data is often such a garden. Our calibrations are still useful at the focused questions at each fork within the garden, but will they transfer successfully to the view from the outside?

I have identified a site for the eighth pillar, but not said what it is. It is an area where a large number of procedures have been developed with partial answers to some specific questions. The pillar may well exist, but no overall structure has yet attracted the general assent needed for recognition. History suggests that this will not appear easily or in one step. Every living science has its mysteries—Astronomy its dark energy and dark matter; Physics its strings and quantum theory; Computer Science its P-NP puzzle; Mathematics the Riemann Hypothesis. The existing seven pillars can support at least partial answers for even the most difficult cases. Statistics is a living science; the support of the seven is strong. We enter a challenging era with strong allies in other fields and great expectations of being equal to the challenge.

NOTES

INTRODUCTION

1. Lawrence (1926). Oddly, Lawrence's only mention of seven pillars is in his title.
2. Herschel (1831), 156; his emphasis.
3. Wilson (1927).
4. Greenfield (1985).

1. AGGREGATION

1. Jevons (1869).
2. Borges ([1942] 1998), 131–137.
3. Gilbert ([1600] 1958), 240–241.
4. Gellibrand (1635).
5. Borough (1581).
6. Gellibrand (1635), 16.
7. Gellibrand (1635).
8. D. B. (1668); his emphasis.

9. Englund (1998), 63.
10. Eisenhart (1974).
11. Patwardhan (2001), 83.
12. Thucydides (1982), 155–156.
13. Köbel (1522).
14. Stigler (1986a), 169–74.
15. Bernard ([1865] 1957), 134–135.
16. Galton (1879, 1883).
17. Galton (1883), 350.
18. Stigler (1984).
19. Stigler (1986a), 39–50.
20. Boscovich (1757).
21. Legendre (1805); Stigler (1986a), 55–61.

2. INFORMATION

1. Galen ([ca. 150 CE] 1944).
2. Stigler (1999), 383–402.
3. Stigler (1986), 70–77; Bellhouse (2011).
4. De Moivre (1738), 239.
5. Laplace (1810).
6. Bernoulli ([1713] 2006); Stigler (2014).
7. Airy (1861), 59–71.
8. Peirce (1879), 197.
9. Ibid., 201.
10. Venn (1878); see also Venn (1888), 495–498.
11. Stigler (1980).
12. Stigler and Wagner (1987, 1988).
13. Cox (2006), 97.

3. LIKELIHOOD

1. Byass, Kahn, and Ivarsson (2011).
2. Arbuthnot (1710).

3. Ibid., 188.
4. Bernoulli (1735).
5. Fisher (1956), 39.
6. Hume (1748).
7. Ibid., 180.
8. Bayes (1764); Stigler (2013).
9. Price (1767).
10. Stigler (1986a), 154.
11. Gavarret (1840).
12. Newcomb (1860a, 1860b).
13. Stigler (2007).
14. Fisher (1922); Stigler (2005); Edwards (1992).
15. Bartlett (1937); Stigler (2007).
16. Neyman and Scott (1948).

4. INTERCOMPARISON

1. Galton (1875), 34.
2. Galton (1869).
3. Gosset (1905), 12.
4. Gosset (1908); Zabell (2008).
5. Gosset (1908), 12.
6. Gosset (1908), 21.
7. Boring (1919).
8. Pearson (1914).
9. Fisher (1915).
10. Fisher (1925).
11. Edgeworth (1885).
12. Efron (1979).
13. Stigler (1999), 77.
14. Jevons (1882).
15. Stigler (1999), 78.
16. Galton (1863), 5.
17. Yule (1926).

5. REGRESSION

1. Charles Darwin to William Darwin Fox, May 23, 1855 (Darwin [1887] 1959, 1:411).
2. Stigler (2012).
3. Galton (1885, 1886).
4. Stigler (1989).
5. Stigler (2010).
6. Darwin (1859).
7. Jenkin (1867); Morris (1994).
8. Galton (1877).
9. Stigler (1999), 203–238.
10. Galton (1889).
11. Hanley (2004).
12. Galton (1885, 1886).
13. Galton (1886).
14. Galton (1889), 109.
15. Fisher (1918).
16. Friedman (1957).
17. Galton (1888).
18. Stigler (1986a), 315–325, 342–358.
19. Dale (1999); Zabell (2005).
20. Bayes (1764).
21. Stigler (1986b).
22. Stigler (1986a), 126.
23. Stigler (1986b).
24. Secrist (1933), i.
25. Stigler (1990).
26. Berkeley (1710).
27. Pearson, Lee, and Bramley-Moore (1899), 278.
28. Hill (1965).
29. Newcomb (1886), 36.
30. Wright (1917, 1975).
31. Goldberger (1972); Kang and Seneta (1980); Shafer (1996).

32. Herschel (1857), 436, in a review first published in 1850 of an 1846 book by Quetelet.
33. Talfourd (1859).

6. DESIGN

1. Dan. 1:8–16.
2. Crombie (1952), 89–90.
3. Jevons (1874), 2:30.
4. Fisher (1926), 511.
5. Young (1770), 2 (2nd Div.): 268–306.
6. Edgeworth (1885), 639–640.
7. Bortkiewicz (1898).
8. Cox (2006), 192.
9. Peirce (1957), 217.
10. Stigler (1986a), 254–261.
11. Peirce and Jastrow (1885).
12. Fisher (1935), ch. V; Box (1978), 127–128.
13. Fisher (1939), 7.
14. Kruskal and Mosteller (1980).
15. Fienberg and Tanur (1996).
16. Neyman (1934), 616.
17. Matthews (1995); Senn (2003).
18. Stigler (2003).

7. RESIDUAL

1. Herschel (1831), 156.
2. Mill (1843), 1:465.
3. Stigler (1986a), 25–39.
4. Yule (1899); Stigler (1986a), 345–358.
5. Fisher (1922); Stigler (2005).
6. Present in Neyman and Pearson (1933), although not referred to as a "lemma" until Neyman and Pearson (1936).
7. Pearson (1900).

8. Labby (2009). See also www.youtube.com/, search for "Weldon's Dice Automated."
9. Stigler (2008).
10. Labby (2009).
11. Cox (1972).
12. Johnson and Raven (1973).
13. Farr (1852).

CONCLUSION

1. Cox (1965).
2. Marshall (1885), 167–168.
3. Borges ([1941] 1998), 119–128.

REFERENCES

Adrain, Robert (1808). Research concerning the probabilities of the errors which happen in making observations, etc. *The Analyst; or Mathematical Museum* 1(4): 93–109.

Airy, George B. (1861). *On the Algebraical and Numerical Theory of Errors of Observations and the Combination of Observations.* 2nd ed. 1875, 3rd ed. 1879. Cambridge: Macmillan.

Arbuthnot, John (1710). An argument for Divine Providence, taken from the constant regularity observ'd in the births of both sexes. *Philosophical Transactions of the Royal Society of London* 27: 186–190.

Bartlett, Maurice (1937). Properties of sufficiency and statistical tests. *Proceedings of the Royal Society of London* 160: 268–282.

Bayes, Thomas (1764). An essay towards solving a problem in the doctrine of chances. *Philosophical Transactions of the Royal Society of London* 53: 370–418. Offprint unchanged, but with title page giving the title as

"A method of calculating the exact probability of all conclusions founded on induction."

Bellhouse, David R. (2011). *Abraham de Moivre: Setting the Stage for Classical Probability and Its Applications.* Boca Raton, FL: CRC Press.

Berkeley, George (1710). *A Treatise Concerning the Principles of Human Knowledge, Part I.* Dublin: Jeremy Pepyat.

Bernard, Claude ([1865] 1957). *An Introduction to the Study of Experimental Medicine.* Translated into English by Henry Copley Greene, 1927. Reprint, New York: Dover.

Bernoulli, Daniel (1735). Recherches physiques et astronomiques sur le problème proposé pour la seconde fois par l'Académie Royale des Sciences de Paris. Quelle est la cause physique de l'inclinaison de plans des orbites des planètes par rapport au plan de l'équateur de la révolution du soleil autour de son axe; Et d'où vient que les inclinaisons de ces orbites sont différentes entre elles. In *Die Werke von Daniel Bernoulli: Band 3, Mechanik,* 241–326. Basel: Birkhäuser, 1987.

Bernoulli, Daniel (1769). *Dijudicatio maxime probabilis plurium observationum discrepantium atque verisimillima inductio inde formanda.* Manuscript. Bernoulli MSS f. 299–305, University of Basel.

Bernoulli, Jacob ([1713] 2006). *Ars Conjectandi.* Translated into English with introductory notes by Edith Dudley Sylla as *The Art of Conjecturing.* Baltimore: Johns Hopkins University Press.

Borges, Jorge Luis ([1941, 1942] 1998). *Collected Fictions.* Trans. Andrew Hurley. New York: Penguin.

Boring, Edwin G. (1919). Mathematical vs. scientific significance. *Psychological Bulletin* 16: 335–338.

Borough, William (1581). *A Discours of the Variation of the Cumpas, or Magneticall Needle.* In Norman (1581).

Bortkiewicz, Ladislaus von (1898). *Das Gesetz der kleinen Zahlen*. Leipzig: Teubner.

Boscovich, Roger Joseph (1757). De litteraria expeditione per pontificiam ditionem. *Bononiensi Scientiarum et Artium Instituto atque Academia Commentarii* 4: 353–396.

Box, Joan Fisher (1978). *R. A. Fisher: The Life of a Scientist*. New York: Wiley.

Bravais, Auguste (1846). Analyse mathématique sur les probabilités des erreurs de situation d'un point. *Mémoires présents par divers savants à l'Académie des Sciences de l'Institut de France: Sciences mathématiques et physiques* 9: 255–332.

Byass, Peter, Kathleen Kahn, and Anneli Ivarsson (2011). The global burden of coeliac disease. *PLoS ONE* 6: e22774.

Colvin, Sidney, and J. A. Ewing (1887). *Papers Literary, Scientific, &c. by the late Fleeming Jenkin, F.R.S., LL.D.; With a Memoir by Robert Louis Stevenson*. 2 vols. London: Longmans, Green, and Co.

Cox, David R. (1965). A remark on multiple comparison methods. *Technometrics* 7: 223–224.

Cox, David R. (1972). Regression models and life tables. *Journal of the Royal Statistical Society Series B* 34: 187–220.

Cox, David R. (2006). *Principles of Statistical Inference*. Cambridge: Cambridge University Press.

Crombie, Alistair C. (1952). Avicenna on medieval scientific tradition. In *Avicenna: Scientist and Philosopher, a Millenary Symposium*, ed. G. M. Wickens. London: Luzac and Co.

D. B. (1668). An extract of a letter, written by D. B. to the publisher, concerning the present declination of the magnetick needle, and the tydes. *Philosophical Transactions of the Royal Society of London* 3: 726–727.

Dale, Andrew I. (1999). *A History of Inverse Probability.* 2nd ed. New York: Springer.

Darwin, Charles R. (1859). *The Origin of Species by Means of Natural Selection, or The Preservation of Favored Races in the Struggle for Life.* London: John Murray.

Darwin, Francis, ed. ([1887] 1959). *The Life and Letters of Charles Darwin.* 2 vols. New York: Basic Books.

De Moivre, Abraham (1738). *The Doctrine of Chances.* 2nd ed. London: Woodfall.

Didion, Isidore (1858). *Calcul des probabilités appliqué au tir des projectiles.* Paris: J. Dumaine et Mallet-Bachelier.

Edgeworth, Francis Ysidro (1885). On methods of ascertaining variations in the rate of births, deaths and marriages. *Journal of the [Royal] Statistical Society* 48: 628–649.

Edwards, Anthony W. F. (1992). *Likelihood.* Exp. ed. Cambridge: Cambridge University Press.

Efron, Bradley (1979). Bootstrap methods: Another look at the jackknife. *Annals of Statistics* 7: 1–26.

Eisenhart, Churchill (1974). The development of the concept of the best mean of a set of measurements from antiquity to the present day. 1971 A.S.A. Presidential Address. Unpublished. http://galton.uchicago.edu/~stigler/eisenhart.pdf.

Englund, Robert K. (1998). Texts from the late Uruk period. In Josef Bauer, Robert K. Englund, and Manfred Krebernik, *Mesopotamien: Späturuk-Zeit und Frühdynastische Zeit,* Orbis Biblicus et Orientalis 160/1, 15–233. Freiburg: Universitätsverlag.

Farr, William (1852). *Report on the Mortality of Cholera in England, 1848–49.* London: W. Clowes and Sons.

Fienberg, Stephen E., and Judith M. Tanur (1996). Reconsidering the fundamental contributions of Fisher and Neyman on experimentation and sampling. *International Statistical Review* 64: 237–253.

Fisher, Ronald A. (1915). Frequency distribution of the values of the correlation coefficient in samples from an indefinitely large population. *Biometrika* 10: 507–521.

Fisher, Ronald A. (1918). The correlation between relatives on the supposition of Mendelian inheritance. *Philosophical Transactions of the Royal Society of Edinburgh* 52: 399–433.

Fisher, Ronald A. (1922). On the mathematical foundations of theoretical statistics. *Philosophical Transactions of the Royal Society of London* Series A 222: 309–368.

Fisher, Ronald A. (1925). *Statistical Methods for Research Workers*. Edinburgh: Oliver and Boyd.

Fisher, Ronald A. (1926). The arrangement of field trials. *Journal of Ministry of Agriculture* 33: 503–513.

Fisher, Ronald A. (1935). *The Design of Experiments*. Edinburgh: Oliver and Boyd.

Fisher, Ronald A. (1939). "Student." *Annals of Eugenics* 9: 1–9.

Fisher, Ronald A. (1956). *Statistical Methods and Scientific Inference*. Edinburgh: Oliver and Boyd.

Friedman, Milton (1957). *A Theory of the Consumption Function*. Princeton, NJ: Princeton University Press.

Galen ([ca. 150 CE] 1944). *Galen on Medical Experience*. First edition of the Arabic version, with English translation and notes by R. Walzer. Oxford: Oxford University Press.

Galton, Francis (1863). *Meteorographica, or Methods of Mapping the Weather*. London: Macmillan.

Galton, Francis (1869). *Hereditary Genius: An Inquiry into Its Laws and Consequences.* London: Macmillan.

Galton, Francis (1875). Statistics by intercomparison, with remarks on the law of frequency of error. *Philosophical Magazine* 4th ser. 49: 33–46.

Galton, Francis (1877). Typical laws of heredity. *Proceedings of the Royal Institution of Great Britain* 8: 282–301.

Galton, Francis (1879). Generic images. *Proceedings of the Royal Institution of Great Britain* 9: 161–170.

Galton, Francis (1883). *Inquiries into Human Faculty, and Its Development.* London: Macmillan.

Galton, Francis (1885). Opening address as president of the anthropology section of the B.A.A.S., September 10, 1885, at Aberdeen. *Nature* 32: 507–510; *Science* (published as "Types and their inheritance") 6: 268–274.

Galton, Francis (1886). Regression towards mediocrity in hereditary stature. *Journal of the Anthropological Institute of Great Britain and Ireland* 15: 246–263.

Galton, Francis (1888). Co-relations and their measurement, chiefly from anthropological data. *Proceedings of the Royal Society of London* 45: 135–145.

Galton, Francis (1889). *Natural Inheritance.* London: Macmillan.

Gavarret, Jules (1840). *Principes généraux de statistique médicale.* Paris: Béchet jeune et Labé.

Gellibrand, Henry (1635). *A Discourse Mathematical on the Variation of the Magneticall Needle.* London: William Jones.

Gilbert, William ([1600] 1958). *De Magnete.* London: Peter Short. Reprint of English translation, New York: Dover.

Goldberger, Arthur S. (1972). Structural equation methods in the social sciences. *Econometrica* 40: 979–1001.

Gosset, William Sealy (1905). The application of the "law of error" to the work of the brewery. *Guinness Laboratory Report* 8(1). (Brewhouse report, November 3, 1904; with board endorsement, March 9, 1905.)

Gosset, William Sealy (1908). The probable error of a mean. *Biometrika* 6: 1–24.

Greenfield, Jonas C. (1985). The Seven Pillars of Wisdom (Prov. 9:1): A mistranslation. *The Jewish Quarterly Review*, new ser., 76(1): 13–20.

Hanley, James A. (2004). "Transmuting" women into men: Galton's family data on human stature. *American Statistician* 58: 237–243.

Herschel, John (1831). *A Preliminary Discourse on the Study of Natural Philosophy.* London: Longman et al.

Herschel, John (1857). *Essays from the Edinburgh and Quarterly Reviews.* London: Longman et al.

Hill, Austin Bradford (1965). The environment and disease: Association or causation? *Proceedings of the Royal Society of Medicine* 58: 295–300.

Hume, David (1748). Of miracles. In *Philosophical Essays Concerning Human Understanding*, essay 10. London: Millar.

Hutton, Charles (ca. 1825). *A Complete Treatise on Practical Arithmetic and Book-Keeping, Both by Single and Double Entry, Adapted to Use of Schools.* New ed., n.d., corrected and enlarged by Alexander Ingram. Edinburgh: William Coke, Oliver and Boyd.

Jenkin, Fleeming (1867). Darwin and *The Origin of Species. North British Review*, June 1867. In Colvin and Ewing (1887), 215–263.

Jevons, William Stanley (1869). The depreciation of gold. *Journal of the Royal Statistical Society* 32: 445–449.

Jevons, W. Stanley (1874). *The Principles of Science: A Treatise on Logic and Scientific Method.* 2 vols. London: Macmillan.

Jevons, William Stanley (1882). The solar-commercial cycle. *Nature* 26: 226–228.

Johnson, Michael P., and Peter H. Raven (1973). Species number and endemism: The Galápagos Archipelago revisited. *Science* 179: 893–895.

Kang, Kathy, and Eugene Seneta (1980). Path analysis: An exposition. *Developments in Statistics* (P. Krishnaiah, ed.) 3: 217–246.

Köbel, Jacob (1522). *Von Ursprung der Teilung.* Oppenheym.

Kruskal, William H., and Frederick Mosteller (1980). Representative sampling IV: The history of the concept in statistics, 1895–1939. *International Statistical Review* 48: 169–195.

Labby, Zacariah (2009). Weldon's dice, automated. *Chance* 22(4): 6–13.

Lagrange, Joseph-Louis. (1776). Mémoire sur l'utilité de la méthode de prendre le milieu entre les résultats de plusieurs observations; dans lequel on examine les avantages de cette méthode par le calcul des probabilités, & où l'on résoud différents problêmes relatifs à cette matière. *Miscellanea Taurinensia* 5: 167–232.

Lambert, Johann Heinrich (1760). *Photometria, sive de Mensura et Gradibus Luminis, Colorum et Umbrae.* Augsburg, Germany: Detleffsen.

Laplace, Pierre Simon (1774). Mémoire sur la probabilité des causes par les évènements. *Mémoires de mathématique et de physique, présentés à l'Académie Royale des Sciences, par divers savans, & lû dans ses assemblées* 6: 621–656. Translated in Stigler (1986b).

Laplace, Pierre Simon (1810). Mémoire sur les approximations des formules qui sont fonctions de très-grands nombres, et sur leur application aux probabilités. *Mémoires de la classe des sciences mathématiques et physiques de l'Institut de France* Année 1809: 353–415, Supplément 559–565.

Laplace, Pierre Simon (1812). *Théorie analytique des probabilités.* Paris: Courcier.

Lawrence, T. R. (1926). *Seven Pillars of Wisdom.* London.

Legendre, Adrien-Marie (1805). *Nouvelles méthodes pour la détermination des orbites des comètes.* Paris: Firmin Didot.

Loterie (An IX). *Instruction à l'usage des receveurs de la Loterie Nationale, établis dans les communes de départements.* Paris: L'Imprimerie Impériale.

Maire, Christopher, and Roger Joseph Boscovich (1770). *Voyage Astronomique et Géographique, dans l'État de l'Église.* Paris: Tilliard.

Marshall, Alfred (1885). The present position of economics. In *Memorials of Alfred Marshall*, ed. A. C. Pigou, 152–174. London: Macmillan, 1925.

Matthews, J. Rosser (1995). *Quantification and the Quest for Medical Certainty.* Princeton, NJ: Princeton University Press.

Mill, John Stuart (1843). *A System of Logic, Ratiocinative and Inductive.* 2 vols. London: John W. Parker.

Morris, Susan W. (1994). Fleeming Jenkin and *The Origin of Species*: A reassessment. *British Journal for the History of Science* 27: 313–343.

Newcomb, Simon (1860a). Notes on the theory of probabilities. *Mathematical Monthly* 2: 134–140.

Newcomb, Simon (1860b). On the objections raised by Mr. Mill and others against Laplace's presentation of the doctrine of probabilities. *Proceedings of the American Academy of Arts and Sciences* 4: 433–440.

Newcomb, Simon (1886). *Principles of Political Economy.* New York: Harper and Brothers.

Neyman, Jerzy (1934). On two different aspects of the representative method. *Journal of the Royal Statistical Society* 97: 558–625.

Neyman, Jerzy, and Egon S. Pearson (1933). On the problem of the most efficient tests of statistical hypotheses. *Philosophical Transactions of the Royal Society of London* Series A 231: 289–337.

Neyman, Jerzy, and Egon S. Pearson (1936). Contributions to the theory of testing statistical hypotheses: I. Unbiassed critical regions of Type A and Type A₁. *Statistical Research Memoirs* (ed. J. Neyman and E. S. Pearson) 1: 1–37.

Neyman, Jerzy, and Elizabeth L. Scott (1948). Consistent estimates based on partially consistent observations. *Econometrica* 16: 1–32.

Nightingale, Florence (1859). *A Contribution to the Sanitary History of the British Army during the Late War with Russia*. London: John W. Parker.

Norman, Robert (1581). *The Newe Attractiue*. London: Richard Ballard.

Patwardhan, K. S., S. A. Naimpally, and S. L. Singh (2001). *Lilavati of Bhaskaracarya*. Delhi: Motilal Banarsidass.

Pearson, Karl (1900). On the criterion that a given system of deviations from the probable in the case of a correlated system of variables is such that it can be reasonably supposed to have arisen from random sampling. *Philosophical Magazine* 5th ser.50: 157–175.

Pearson, Karl, ed. (1914). *Tables for Statisticians and Biometricians*. Cambridge: Cambridge University Press.

Pearson, Karl, Alice Lee, and Leslie Bramley-Moore (1899). Mathematical contributions to the theory of evolution VI. Genetic (reproductive) selection: Inheritance of fertility in man, and of fecundity in thoroughbred racehorses. *Philosophical Transactions of the Royal Society of London* Series A 192: 257–330.

Peirce, Charles S. (1879). Note on the theory of the economy of research. Appendix 14 of *Report of the Superintendent of the United States Coast Survey* [for the year ending June 1876]. Washington, DC: GPO.

Peirce, Charles S. (1957). *Essays in the Philosophy of Science*. Ed. V. Tomas. Indianapolis: Bobbs-Merrill.

Peirce, Charles S., and Joseph Jastrow (1885). On small differences of sensation. *Memoirs of the National Academy of Sciences* 3: 75–83.

Playfair, William (1801). *The Statistical Breviary*. London: T. Bensley.

Price, Richard (1767). *Four Dissertations*. London: Millar and Cadell.

Pumpelly, Raphael (1885). Composite portraits of members of the National Academy of Sciences. *Science* 5: 378–379.

Secrist, Horace (1933). *The Triumph of Mediocrity in Business*. Evanston, IL: Bureau of Business Research, Northwestern University.

Senn, Stephen (2003). *Dicing with Death: Chance, Risk and Health*. Cambridge: Cambridge University Press.

Shafer, Glenn (1996). *The Art of Causal Conjecture*. Appendix G, 453–478. Cambridge, MA: MIT Press.

Simpson, Thomas (1757). An attempt to shew the advantage arising by taking the mean of a number of observations, in practical astronomy. In *Miscellaneous Tracts*, 64–75 and plate. London: Nourse Press.

Stigler, Stephen M. (1980). An Edgeworth curiosum. *Annals of Statistics* 8: 931–934.

Stigler, Stephen M. (1984). Can you identify these mathematicians? *Mathematical Intelligencer* 6(4): 72.

Stigler, Stephen M. (1986a). *The History of Statistics: The Measurement of Uncertainty before 1900*. Cambridge, MA: Harvard University Press.

Stigler, Stephen M. (1986b). Laplace's 1774 memoir on inverse probability. *Statistical Science* 1: 359–378.

Stigler, Stephen M. (1989). Francis Galton's account of the invention of correlation. *Statistical Science* 4: 73–86.

Stigler, Stephen M. (1990). The 1988 Neyman Memorial Lecture: A Galtonian perspective on shrinkage estimators. *Statistical Science* 5: 147–155.

Stigler, Stephen M. (1999). *Statistics on the Table*. Cambridge, MA: Harvard University Press.

Stigler, Stephen M. (2003). Casanova, "Bonaparte," and the Loterie de France. *Journal de la Société Française de Statistique* 144: 5–34.

Stigler, Stephen M. (2005). Fisher in 1921. *Statistical Science* 20: 32–49.

Stigler, Stephen M. (2007). The epic story of maximum likelihood. *Statistical Science* 22: 598–620.

Stigler, Stephen M. (2008). Karl Pearson's theoretical errors and the advances they inspired. *Statistical Science* 23: 261–271.

Stigler, Stephen M. (2010). Darwin, Galton, and the statistical enlightenment. *Journal of the Royal Statistical Society* Series A 173: 469–482.

Stigler, Stephen M. (2012). Karl Pearson and the Rule of Three. *Biometrika* 99: 1–14.

Stigler, Stephen M. (2013). The true title of Bayes's essay. *Statistical Science* 28: 283–288.

Stigler, Stephen M. (2014). Soft questions, hard answers: Jacob Bernoulli's probability in historical context. *International Statistical Review* 82: 1–16.

Stigler, Stephen M., and Melissa J. Wagner (1987). A substantial bias in nonparametric tests for periodicity in geophysical data. *Science* 238: 940–945.

Stigler, Stephen M., and Melissa J. Wagner (1988). Testing for periodicity of extinction: Response. *Science* 241: 96–99.

Talfourd, Francis (1859). *The Rule of Three, a Comedietta in One Act.* London: T. H. Levy.

Thucydides (1982). *The History of the Peloponnesian War.* Trans. Richard Crawley. New York: Modern Library.

Venn, John (1878). The foundations of chance. *Princeton Review*, September, 471–510.

Venn, John (1888). *The Logic of Chance.* 3rd ed. London: Macmillan.

Watson, William Patrick (2013). *Catalogue 19: Science, Medicine, Natural History.* London.

Wilson, Edwin B. (1927). What is statistics? *Science* 65: 581–587.

Wright, Sewall (1917). The average correlation within subgroups of a population. *Journal of the Washington Academy of Sciences* 7: 532–535.

Wright, Sewall (1975). Personal letter to Stephen Stigler, April 28.

Young, Arthur (1770). *A Course of Experimental Agriculture.* 2 vols. London: Dodsley.

Yule, G. Udny (1899). An investigation into the causes of changes in pauperism in England, chiefly during the last two intercensal decades, I. *Journal of the Royal Statistical Society* 62: 249–295.

Yule, G. Udny (1926). Why do we sometimes get nonsense-correlations between time-series? *Journal of the Royal Statistical Society* 89: 1–96.

Zabell, Sandy L. (2005). *Symmetry and Its Discontents: Essays on the History of Inductive Philosophy.* Cambridge: Cambridge University Press.

Zabell, Sandy L. (2008). On Student's 1908 article "The Probable Error of a Mean." *Journal of the American Statistical Association* 103: 1–20.

ACKNOWLEDGMENTS

This book was originally conceived a decade ago but has taken much longer to finish than expected. The delay has provided an opportunity to explore new areas and revisit others with a different focus. Some of the ideas within the book have been presented over the years in a number of talks, and I have benefited from constructive comments from many people, including David Bellhouse, Bernard Bru, David R. Cox, Persi Diaconis, Amartya K. Dutta, Brad Efron, Michael Friendly, Robin Gong, Marc Hallin, Peter McCullagh, Xiao-Li Meng, Victor Paneretos, Robert Richards, Eugene Seneta, Michael Stein, Chris Woods, and Sandy Zabell. I am grateful to Nat Schenker for his invitation to give the American

Statistical Association President's Invited Lecture in August 2014, where I converged on the present framework. Even so, this project would never have been completed without the encouragement of my patient editor, Michael Aronson, and my impatient wife, Virginia Stigler.

INDEX